Sentiment
from the sacred to the human

004 *Architecture of Sentiment* _ Aldo Vanini

to Death

012 New Crematorium at the Woodland Cemetery _ Johan Celsing Arkitektkontor

034 Communal Crematorium in Ringsted _ Henning Larsen Architects

042 Crematorium in Amiens _ PLAN01

058 Mahaprasthanam Hindu Crematorium and Cemetery _ DA Studios

068 Jagged Concrete Canopy at a Cemetery _ Ron Shenkin Studio

080 Sayama Forest Chapel _ Hiroshi Nakamura & NAP

in memory

096 The Rivesaltes Memorial _ Rudy Ricciotti + Passelac & Roques Architects

116 Mausoleum of the Martyrdom of Polish Villages _ Nizio Design International

126 The Ring of Remembrance _ Agence d'Architecture Philippe Prost

138 Bologna Shoah Memorial _ SET Architects

154 MIT Collier Memorial _ Höweler + Yoon Architecture

162 Stone Memorial in Ishinomaki _ Koishikawa architects

172 Memorial at Utøya _ 3RW Arkitekter

184 Index

建筑情感

Sentim

从宗教到世俗
from the sacred to the human

人们对于死亡自始至终都心生恐惧，因此，为了消除这种恐惧，人们就有了为逝者建造神圣之地的想法。

最近，在一篇文章中，尼尔森·莫塔引用了阿道夫·路斯的一些话语，认为坟墓和纪念碑是建筑艺术的精华部分。

实际上，我们可以认为这种说法非常激进，但毫无疑问，撇开单纯的功能不谈，这种建筑纯粹代表了一种情感和记忆，是表达神圣之地的自然方式。坟墓或火葬场都属于这一类型建筑，它们不仅是掩埋或者火化尸体的地方，也是一种精心制作的人类艺术品，活着的人借此来表达对逝者的怀念，当然也表达了人们的恐惧，害怕死亡再次降临。在这方面的代表人物是克洛德·列维·斯特劳斯。

The fear of death has haunted mankind since its beginning, and on this fear, on the need to exorcise it, the idea of the sacred has been built.

Recently, in these very pages, Nelson Mota quoted Adolf Loos's words about tombs and monuments being the quintessential part of Architecture as Art.

Actually, we can consider that statement quite radical but, nevertheless, there is no doubt that subtracting the mere functionality from an artifact places it in a sphere of pure representation of sentiment and memory, a sphere that is the natural ground of the expression of the sacred. Belonging to this sphere, the grave or the crematorium are not just instruments to get rid of corpses, but they are elaborate anthropological artifacts by which the living express their respect for the deceased and the fear, well represented by Claude Lévy Strauss, that they

建筑立场系列丛书 No.65

C3

建筑情感：从宗教到世俗
Sentiment
from the sacred to the human

亨宁·拉森建筑事务所等 | 编
杜丹 于风军 孙探春 王京 林英玉 徐雨晨 | 译

大连理工大学出版社

建筑情感：
从宗教到世俗

004 建筑情感 _ Aldo Vanini

关于死亡

012 林中墓地新火葬场 _ Johan Celsing Arkitektkontor

034 灵斯泰兹公共火葬场 _ Henning Larsen Architects

042 亚眠火葬场 _ PLAN01

058 Mahaprasthanam印度教火葬场和公墓 _ DA Studios

068 公墓锯齿状混凝土顶棚 _ Ron Shenkin Studio

080 狭山森林小教堂 _ Hiroshi Nakamura & NAP

关于回忆

096 里韦萨尔特纪念馆 _ Rudy Ricciotti + Passelac & Roques Architects

116 波兰村庄殉难者陵墓 _ Nizio Design International

126 回忆之环 _ Agence d'Architecture Philippe Prost

138 博洛尼亚大屠杀纪念碑 _ SET Architects

154 麻省理工科利尔纪念碑 _ Höweler + Yoon Architecture

162 日本石卷市石头纪念碑 _ Koishikawa architects

172 乌托亚纪念碑 _ 3RW Arkitekter

184 建筑师索引

林中墓地新火葬场_New Crematorium at the Woodland Cemetery/Johan Celsing Arkitektkontor
灵斯泰兹公共火葬场_Communal Crematorium in Ringsted/Henning Larsen Architects
亚眠火葬场_Crematorium in Amiens/PLAN 01
Mahaprasthanam印度教火葬场和公墓_Mahaprasthanam Hindu Crematorium and Cemetery/DA Studios
公墓锯齿状混凝土顶棚_Jagged Concrete Canopy at a Cemetery/Ron Shenkin studio
狭山森林小教堂_Sayama Forest Chapel/Hiroshi Nakamura & NAP
里韦萨尔特纪念馆_The Rivesaltes Memorial/Rudy Ricciotti + Passelac & Roques Architects
波兰村庄殉难者陵墓_Mausoleum of the Martyrdom of Polish Villages/Nizio Design International
回忆之环_The Ring of Remembrance/Agence d'Architecture Philippe Prost
博洛尼亚大屠杀纪念碑_Bologna Shoah Memorial/SET Architects
麻省理工科利尔纪念碑_MIT Collier Memorial/Höweler + Yoon Architecture
日本石卷市石头纪念碑_Stone Memorial in Ishinomaki/Koishikawa Architects
乌托亚纪念碑_Memorial at Utøya/3RW Arkitekter

建筑情感_Architecture of Sentiment/Aldo Vanini

一旦通过火化或者埋葬的仪式来驱除对死亡的恐惧，人们对死亡的特别的情感状态就值得被记忆。通过记忆，这样的时刻总能表达一种集体价值观，让社区更具凝聚力。

然而，现代性引入了一种激进的思潮，关注点从超然的神圣转移到了人文主义的神圣之上，前者是为了驱除对死亡的恐惧，而后者是为了表达对逝者及其亲人的同情。

本书所列的这些项目由于有了人类的虔诚而具有了生气。建筑师们选择了用简洁的建筑词汇来表达那种安详而充满敬意的沉默，同时摒弃了那些与宗教有关的，或者说任何与超然的观点有关的、传统的、陈旧老套的建筑符号。

might come back.
Once the fear is exorcised by means of incineration or burial rites, extraordinary, emotional circumstances of death deserve to be remembered as moments in which, through memory, a community expresses and tightens around its collective values.
Modernity, however, has introduced a radical revolution, shifting the focus from a metaphysical sacredness to a humanistic sacredness, motivated no longer by the exorcism of the fear of death but by the compassion for the deceased and for his loved ones.
The submitted projects are animated by human pietas, opting for a lean architectural vocabulary addressed to a composed and respectful silence, and rejecting the traditional and stereotypical symbols related to religion or, in any case, to a metaphysical view.

建筑情感

在最初的人类自我意识形成的一千年中，人类第一个详尽阐述的超自然的想法很可能就是从有到无转换的问题，古生物学的研究结果似乎也证明了这一点。这一问题甚至早于人类对创世纪和有关神学的关注。

打猎是人们谋生的手段，与打猎时举行的安抚仪式一道，人们用仪式来驱除对死亡的恐惧。这种抽象的恐惧与生命的终结有关，人们担心那些逝去的人可能会阴魂不散。因此，人们创造了专门用于生命终结的地方，这甚至要早于卫生管理要求的出台。

由于其原始的起源，这些专门用来表达对逝者崇拜和尊敬的建筑符号一直都是建筑的基本原型。这就是为什么会将坟墓和纪念碑列为艺术的精粹，如尼尔森·莫塔[1]曾在《C3建筑立场系列丛书》中所引述的那样。

阿道夫·路斯在下面句子中同样充满诗意地表达了逝者祭祀仪式的物质符号和人类意识之间直接而深刻的关系。

"当我们在树林里发现用铁锹堆起的6ft (1.8m) 长、3ft (0.9m) 宽的锥形土堆时，我们会变得很严肃，心里会想：某人葬在这里。这就是建筑。"[2]

本着这样的想法，在阿道夫·路斯给自己设计的坟墓中，没有添加任何本质以外的东西，都是自然呈现。

在牧师、教堂和君主们所控制的宗教中，本能的精神情感被改变，原来那些正式的、极简的建筑基本原型逐渐被赋予了更多超自然因素。坟墓和陵墓逐渐成为强大的宗教和政治的宣传机器，将超自然的力量和权力联系起来。

然而，通过建筑，人们缅怀的不仅是逝者的遗体，还有对逝者的

Architecture of Sentiment

It is likely – and paleontological findings seem to prove it – that the first metaphysical thought elaborated by humankind, in the first thousand years of the formation of self-consciousness, was the question of the transition from being and non-being: an agonizing question that also preceded the one concerning the creation and the related divinity.

Together with propitiatory rites for hunting – and then to ensure his own survival through food–Man conceived rituals to exorcise the fear of death. The abstract terror connected to the end of life, and to a possible return of those who had already abandoned it – even before hygienic necessities – led to the creation of places dedicated to the existential termination.

Because of their primordial origin, the signs of artifacts dedicated to the worship and to the respect of the dead are fundamental archetypes for the architecture of all time. This is the reason why Adolf Loos, as quoted in a recent issue of this magazine by Nelson Mota[1], attributed almost exclusively to tombs and monuments the quality of Art.

The same Adolf Loos poetically set, in one sentence, this immediate and deep relationship between the material signs of the cult of the dead and human consciousness.

"When we find a mound in the woods, six feet long and three feet wide, raised to a pyramidal form by means of a spade, we become serious and something in us says: someone was buried here. That is Architecture."[2]

Following this concept, in his self-designed grave, Adolf Loos didn't add anything to that archetypical essentiality.

The transformation of the instinctive spiritual feeling in a religion administered by clerics, churches and monarchs has progressively loaded those formal and minimalist archetypes with ever richer suprasegmental elements. Tombs and mausoleums have been progressively embellished to become formidable religious and political propaganda machines, mediating bridges between the metaphysical and the Power.

However, humankind didn't celebrate by means of architec-

狭山森林小教堂，日本
Sayama Forest Chapel, Japan

记忆，或对重大事件的记忆，尤其是通过将这些记忆与社区的集体情感联系起来，来提高或加强人们的认同感。

只有到了19世纪，随着启蒙运动和现代科学的出现，死亡和仪式之间的关系才逐渐褪去了神秘的宗教色彩。

在欧洲，拿破仑·波拿巴统治时期，强制实行根据卫生管理条例处理尸体，从此，死亡的神秘性更多地具有了人类同情色彩，减少了其宗教色彩。此外，综观整个仪式的风格特征，它越来越倾向于功能理性主义，并且逐渐抛弃了各种形式的修饰主义。在其后几十年里，世界大战带来的恐慌导致人们本能地抵制那些成为各种冲突的宣传工具的华丽修饰。因此，即使是对逝者寄予的哀思也失去了曾经所有的那种传统的华丽修饰痕迹。

最终，基于人类团结一致的原则，基于尘世色彩的虔诚，形成了一种新的修饰方式。

新修饰方式的主张更加明显地体现在纪念馆的主题上，以前纪念馆主要彰显军功，现在主要用来表达公众对灾难受害者的同情与悼念。

焦点的转变充分体现在建筑上，那就是使用少量但是具有强烈感情色彩的符号表达尊敬、情感和默哀，是一种回到原始起源的人类学回归，用提倡平静地接受"人固有一死"这一结局来解决祖先对于死亡的恐惧。

死亡仪式是在每种文化中都起着核心作用，它的人类学本质将其置于一种对立的体系之中。"生与死"的对立和"城市与农村"的对立也相互关联。传统上，城市是居住生活的地方，外围是农村，是野兽出没和掩埋尸体的地方，这样做不仅是因为卫生原因，也标示着排斥与拒绝。

位于日本崎玉县的狭山森林小教堂，地处林区和墓区之间，与森

ture only the physical remains of the deceased, but also their memory, in particular by connecting it, or the memory of exceptional events, to the collective sentiment of the community, in exaltation or strengthening of their identity.
Only in the nineteenth century, with the advent of the Enlightenment and of modern science, did the relationship between death and its rituals move away from the obscure mystery rigidly administered by the official religious organizations.
In Europe, the stormy passage of Napoleon Bonaparte imposed specific hygiene rules for managing corpses, opening the way to a relationship with the mystery of death oriented more toward human compassion than to religious devotion. Add to that a general review of the stylistic features increasingly oriented toward a functional rationalism and to the progressive abandonment of all forms of decorativism. In the following decades, the horrors of the World Wars led common sense to reject the rhetoric that had been the propaganda apparatus of those conflicts. Consequently, even the feeling addressed to the deceased lost all traces of that traditional rhetoric that had accompanied it.
Eventually, a new form of rhetoric established itself, now founded on principles of human solidarity, on a civil pietas made of an earthlier nature.
The affirmation of this new rhetoric is even more evident in the subject of memorials: in the past mainly dedicated to the celebration of military glory, now dedicated to the compassion for the victims of events that have captured public emotion.
This shift of focus is fully reflected in architectural representation, which uses few and strong signs oriented to respect, emotion and silence, in a sort of anthropological return to the primordial origins, but solving the ancestral terror in favor of a serene acceptance of the ineluctability of the end.
The anthropological nature of the central role played by death rituals in every culture places them in a system of oppositions. The "life-death" opposition is also connected with the "city-countryside" one. Traditionally, the city was the place of the

波兰村庄殉难者陵墓，波兰
Mausoleum of the Martyrdom of Polish Villages, Poland

林静静地交流着。建筑事务所Hiroshi Nakamura & NAP利用Gassho-zukuri这一传统结构形式，即人们合掌祈祷的形状，将生命和祈祷联系起来。

北欧文化中以功能为导向的实用设计方法既很好地处理了死亡的神秘，同时也给予生者诸多关注。丹麦灵斯泰兹公共火葬场由Henning Larsen建筑师事务所设计，其设计想方设法让逝者亲属平复心情，也让在此工作的员工心境平和，同时减少对环境的影响。这一建筑理念体现了著名的丹麦设计原则，使整座建筑的抽象几何形状与组成部分的物质性及装饰的精致优雅相结合。由于巧妙地使用了光线，整体氛围既庄严又宽慰人心，没有了那种与死亡相关的不适感。

事实上，"光与影"的对立和"生与死"的对立密切相关。与Henning Larsen建筑师事务所选择采用强光不同，Johan Celsing选择利用森林的影子，创造了别样的空间，用来激发人们对于生命终结所感受的那种强烈的情绪。位于斯德哥尔摩郊区的林中墓地新火葬场，是一个特别敏感的规划项目，既想要在这一茂密而美丽的林地中占有一席之地，又要加强与贡纳尔·阿斯普伦德设计的林中墓地这一著名建筑的协调，而这一建筑又是20世纪瑞典的重要地标。因此，林中墓地新火葬场的设计师巧妙地通过所使用的材料、门廊和庭院，营造出静谧安详的情感环境，与室外自然环境统一、呼应。

位于法国亚眠的火葬场项目由PGP设计，与上述的火葬场都不同，营造了一个全新的与生命告别的视角。该建筑设计没有采用几何直角，而是采用圆形的空间性来区分于"生者"世界，几乎创造了一种全新的神话叙述方式。正式的设计理念是尽可能远离传统的神圣性，目的在于减少亲人失去逝者的伤痛，为人们提供一个自我阐释死亡这一紧张时刻的空间。

位于以色列Pardesiya的公墓由Ron Shenkin设计，他把我们带回到

living, outside of which—and not only for hygienic reasons, but to mark its exclusion—was the countryside, "place non-place" for beasts and dead.

The Sayama Forest Chapel, in Saitama Prefecture, Japan, silently communicates with the forest, as a meeting point between the city of the dead and nature. Hiroshi Nakamura & NAP resort to the traditional form of the Gassho-zukuri, the shape of two palms put together in prayer, in an ideal link between living and praying.

The pragmatic, functionally oriented approach of Northern European cultures manages the mystery of death, paying the greatest attention to the care of the living. In Ringsted Communal Crematorium, Henning Larsen Architects have managed everything to put not only the relatives, but also the staff in complete peace of mind conditions, and to reduce the environmental impact. The concept responds to the well-known principles of Danish design, combining the abstract geometry of the volumes with the materiality of the components and the refined elegance of the furnishings. The atmosphere is austere and reassuring at the same time, thanks to a clever use of light, to keep out the uneasy feeling of darkness related to death.

In fact, the "light-shadow" opposition is strongly related to the "life-death" one. Differently from Henning Larsen's choice of a bright light, in the New Crematorium at the Woodland Cemetery outside Stockholm, Johan Celsing plays with the shadows of the forest in creating spaces aimed at inducing strong emotions tied to the end of life. It is a particularly sensitive planning program, for the need to reclaim its own space into a thick and beautiful woodland, and for the dialogue with the celebrated architecture of Eric Gunnar Asplund's Cemetery Chapel, a fundamental Swedish landmark of the twentieth century. A clever use of materials, of a porch and of a courtyard, contributes to create an emotional peaceful continuity with the outside natural environment.

The project for the Crematorium in Amiens, by Parisian group Plan01, goes away from all of this, imagining a completely new vision of the celebration of detachment from life. This

公墓锯齿状混凝土顶棚，以色列
Jagged Concrete Canopy at a Cemetery, Israel

最古老的悼念方式。Ron Shenkin设计了一个供参加葬礼的人们等待和聆听悼词的开放空间，这一空间由可以令人联想到逝者回归自然的一些程式化的符号所定义，顶棚下的一切都强化了这种"尘归尘，土归土"的宿命，让所有参加悼念活动的人都感受到这种天地联系。

在印度，在对Mahaprasthanam印度教火葬场和公墓进行重建的项目中，使古老的习俗留存至现在是更加激进的做法，尽管这些古老习俗只体现在现代建筑词汇中。盛大的仪式按照《博伽梵歌》的精神来举行，而用来举行仪式的空间全部用混凝土墙来界定。墙上只简简单单刻有印度教经典诗文，墙体方向不一。在整个项目中，即使是火化这一火葬场最原始的功能，也体现了当代建筑语言。

然而，隐藏或者消除死亡带给人的悲痛并不总是合适的，死亡也并非是生命的自然终结，需要人们平静、顺从地去面对。相反，第二次世界大战期间纳粹占领波兰村庄，将所有波兰村庄人都杀害了，这种惨绝人寰的灭绝行为就要求建筑能将当时关于那种悲剧的记忆和人们心中的愤怒情绪保存下来。由Nizio Design International建筑事务所设计的波兰村庄殉难者陵墓综合设施本身带有强有力的、重要的象征符号。这座陵墓唤起了人们对集体生活的回忆，几乎理想化地再现了一个村庄的重建和毁灭，裸露的混凝土墙造型扭曲，不规则地留有孔洞，用来象征炮击后的结果，取得了惊人的效果。陵墓中将收藏关于大屠杀的文件，其分段设计的布局让人们回忆起那长长的一个村庄又一个村庄遇难者的名单。墙面上有粗糙的木桌的纹理，记录那些地方曾经发生的暴行。在这里，记忆不是为了抚慰人们内心的创伤，而是告诫人们，教育人们。

主要世界历史常常不关注非全球性事件，但是非全球性事件其悲剧性不减。纪念碑或纪念馆就是为了记录这些重大事件，并向这些事件亲历者表达敬意。在内战期间，里韦萨尔特难民营收留了西班牙难

creates almost a new mythological narrative, based on the suggestive idea of abandonment of the orthogonal geometry, distinguishing the world of the living to a spatiality dominated by circular forms. The formal concept is about as far as it can be from a traditional sacredness and is aimed at minimizing the drama of the loss, offering a contemporary space open to an individual interpretation of this intense moment.

In the Cemetery Pavilion in Pardesiya, Israel, Ron Shenkin takes us back to the most ancient ways of condolences. He uses a community gathering and eulogies in an open space defined by stylized signs reminiscent of the return of the dead to nature, everything under a roof that reinforces the sense of the link among the mourners who are brought together in commemoration.

More radical is the survival of ancient customs, though represented by a contemporary architectural vocabulary, in the renewal of the Mahaprasthanam Hindu Crematorium and cemetery in India. Spaces, delimited by concrete walls simply decorated by verses from the Hindu scriptures and oriented in a complex scheme, host the elaborate ritual in the Bhagavad Gita spirit. Even the archetypal functions of incineration find, in the compound, a contemporary architectural expression. Not always, however, is it appropriate to hide or delete the drama of death. Not always death the natural end of life, to be faced with serenity and resignation. The extermination of entire communities of Polish villages during the Nazi occupation in World War II required, on the contrary, an architecture capable of maintaining the memory of that tragedy and the feeling of indignation. The complex of the Mausoleum of the Martyrdom of Polish Villages, designed by Nizio Design International, reports with powerful and essential signs. It evokes the very idea of collective living, almost an ideal reconstruction of a village, and its destruction, distorting the bare concrete walls, randomly perforated as a result of shelling, concurring in a dramatic effect. The layout of the segments that will host the documents of the massacres, recalls long lines of aligned victims. The surface of the walls, marked by

里韦萨尔特纪念馆，法国
The Rivesaltes Memorial, France

民。在1941年和1942年，成为俘虏收容所，关押第二次世界大战时期德国的战俘和通敌者。最后，成为"哈基斯人"（阿尔及利亚战争中站在法国一边为法国陆军服务的一些本地阿尔及利亚穆斯林，在阿尔及利亚战争结束后，哈基斯人被法国政府无情地抛弃）的重新安置中心。鲁迪·里奇奥蒂设计的纪念馆如同陀思妥耶夫斯基的《地下室手记》，一个巨大的空心混凝土方块完全掩埋于地下，任凭原来集中营的废墟见证并述说着曾经住在这里的人们所遭受的苦难。建筑师选择用建筑语言来表达严厉的道德宣言。

在当今信息爆炸的时代里，即使是最近发生的事，为了不被层出不穷的新闻消息淹没而被人们遗忘，也需要纪念碑或纪念馆将其留存下来。2011年在乌托娅发生了悲惨的事件，人们为此修建纪念碑的目的不是为了提醒人们当时发生了什么，而是通过保存对那些倒下的年轻人的记忆将历史净化。这一纪念碑由受害者亲属们集体而建，其中包含了人们的参与，体现了与自然的关系，而不仅仅是用一些简单的物体告诉人们受害者是谁。

今天，我们回顾赞美战争中的荣耀的传统主题，将战士的英雄主义暂时抛开，共同赞美战争双方的受害者，表达他们的悲伤和死亡。位于法国洛雷特圣母院的回忆之环由菲利普·普罗斯特设计，一个大大的椭圆形环体，部分悬浮，用坚硬而贫瘠的混凝土来表现在第一次世界大战中牺牲的所有士兵的名字，没有任何区别对待。这种表现手法彻底地背离了传统的关于英雄和祖国的表现手法。

位于波士顿麻省理工学院校园内的肖恩·科利尔纪念碑形似张开的手掌，寓意肖恩·科利尔警官以身殉职以及为社区服务的精神。肖恩·科利尔警官是在值班时被杀害的。麻省理工学院建筑学院团队在J. Meejin Yoon带领下，用巨大的灰色石块来传达力量和团结的理念。这些巨大的石块利用CAD-CAM技术经过机器加工，摆放方向不同，用其

the texture of rough wooden tables, refers to the brutality of what happened in those places. Here, the memory is made not to pacify, but to admonish and educate.

Often the main world history forgets episodes less global, but no less tragic. It is the task of the memorials to document these events and to pay homage to their protagonists. The Camp de Rivesaltes hosted Spanish refugees during the Civil War, was an internment camp in 1941 and 1942, held German prisoners of war and collaborators and, finally, became a relocation center for Harkis, the Algerian loyalist auxiliaries in the French Army, unfairly abandoned by the French government after the Algerian War. A sort of Dostoevsky's "Notes from Underground" by means of architecture, Rudy Ricciotti's Memorial is a concrete monolith developed below the ground level, leaving to the ruins of the Camp the task of witnessing the misery of all those who were housed there. This choice is set up as a severe ethical manifesto implemented by the language of architecture.

Even the most recent events, overwhelmed by today's hectic and forgetful information flow, must rely on memorials to not be forgotten in the daily succession of news. The goal of the project for the memorial of the tragic events of 2011 in Utøya is not to be a sad reminder of what happened, but to purify it by preserving the memory of the young people who fell. Built by means of collective work, the "Dugnad", of the relatives of the victims, the memorial consists more of actions and relationships with nature than of the simple objects that show the names of the victims.

The conventional theme of the celebration of glory in battle is reviewed today, putting aside the heroism of the fighters, to celebrate together the victims of both opposing parties, united in their time of sorrow and death. The Ring of Remembrance, at Notre-Dame-de-Lorette, France, by Philippe Prost, joined in one large elliptical ring, partly suspended, the names of all those killed in battles of World War I, without distinction, represented by the hardness and poverty of the concrete a radical departure from the traditional rhetoric of the hero and homeland.

博洛尼亚大屠杀纪念碑，意大利
Bologna Shoah Memorial, Italy

空间唤起人们对这一血腥事件的记忆，重申了象征主义和集体回忆之间的紧密联系。

　　石卷市石头纪念碑是专门用来纪念一起自然灾害的遇难者的。在2011年袭击日本的地震中，有1.8万人丧生。石卷市石头纪念碑就是为这些遇难者而建的，用来表达整个民族深深的悲痛。纪念碑设计的灵感来自于空间方向和力的概念，它几乎就像是一个指南针，朝向受灾最严重的地区。这在东方文化中很常见。石卷市石头纪念碑既是纪念碑，也是神社，由小引宽也和石川典贵设计，其形状是一个微型堡垒，由无数的石瓦砌成，每片石瓦代表其中一个遇难者。顶部光亮的钢板反射着附近的樱花树，樱花树每年春天开放，提醒着人们，樱花开放的时候，就是那场地震发生的时候。

　　意大利博洛尼亚大屠杀纪念碑由SET建筑师事务所设计，用以纪念在对纳粹集中营的憎恨中结束生命的不幸者的痛苦。用考顿钢修建的两面高墙在两个月之内就建好了——这种效率隐喻着毁灭机器的效率。这两面高墙既让人想起囚犯所待的狭小隔间，又让人想起公墓的壁龛——事实上，这两者也并非有所不同。在最后这个案例中，纪念碑所表现的巨大的痛苦让人们不会遗忘曾经发生的事情，它是通过观看者穿过狭窄的过道时所感受到的愤怒与不安来表达对受害者的敬意。

The metaphor of a stylized open hand is a reminder, in Boston's MIT campus, of the sacrifice and the spirit of the community service of Officer Sean Collier, killed while on duty. The team of MIT's Department of Architecture, led by J. Meejin Yoon, transforms the concepts of strength and unity in massive gray stone blocks, machined by CAD-CAM, playing with the directions and spatial references of the bloody episode and reasserting the close connection between symbolism and collective memory.

Devoted to the victims of a natural disaster, Ishi-no-kinendo, a memorial for the 18,000 victims of the earthquake that hit Japan in 2011, represents a moment of profound grief for the whole national community. As is common in Eastern cultures, it is inspired by spatial directions and forces, almost a compass oriented to the most affected country's areas. Monument and shrine at the same time, it was conceived by Hiroya Kobiki and Noritaka Ishikawa as a miniature bastion formed by a myriad of stone shingles, one for each of the victims. The shiny steel top reflects the nearby cherry tree, which blooms in the spring, a reminder every year at the season when the disaster struck.

The Shoah memorial in Bologna, Italy, is imagined by SET architects as memory of anguish of the unfortunates who ended their lives in the abomination of the Nazi concentration camps. The two high walls of corten steel were built in just two months – almost a further metaphor of the hectic efficiency of that machine of destruction – and they recall at the same time the cubicles of the prisoners and the niches of a cemetery, as the two were not, in fact, dissimilar. In this last example, the enormity of the pain that the monument shouts, prevents reconciliation, but it entrusts the pietas towards the victims to the indignation and the discomfort of the beholder passing through the narrow passage. *Aldo Vanini*

1. Nelson Mota, *Architecture of Memorial*, C3 No.345, May 2013.
2. Adolf Loos, *Architektur*, Wien 1910.

林中墓地新火葬场
Johan Celsing Arkitektkontor

建筑情感:从宗教到世俗 Architecture of Sentiment – from the sacred to the human

新火葬场位于斯德哥尔摩一片林中墓地中原生林起伏的地带。在火葬场的周围是一大片有着上百年树龄的古松树。这栋紧凑的混凝土结构建筑大约离1940年埃里克·贡纳尔·阿斯普伦德设计的主礼拜堂有150m远。在2009举办的年匿名国际竞赛中，新火葬场的主题是"林中之石"。

新火葬场整体布局和室内特色是设计的重点。紧凑的布局让员工一眼就可以看到建筑整体面貌，同时也减少了对珍贵森林的占用。该建筑的主体结构用了裸露的白色混凝土。这样做的目的是展现建筑物的稳固，同时使室内具有温和、宽厚之感。混凝土由纯白的丹麦白水泥混合以白云石碎石而制成。当模板拆除之后，其表面未经任何处理。这样，墙和天花板都有一种不易被觉察但真实可见的建筑痕迹。

考虑到传声效果，一些室内使用了多孔砖。室内白色釉面能够反射并增强从屋顶洞口和缝隙洒进室内的光线。建筑内的公共空间之一是仪式厅，这里可放置棺材或者骨灰盒，吊唁者可以在此举行仪式。建筑内部有一个露天的中庭，火葬场的员工可以在这里休息，不会打扰或影响到送葬者。建筑立面和屋顶都使用了砖，这样可以使表面的规模比较小，同时也与周围松树的树干相呼应。送葬者和游客沿着树林之间由大型花岗岩石板铺成的小路可以到达火葬场。在公共入口处，有一个大的门廊，吊唁者可以在这里聚集等候或休息，欣赏眼前的自然林地。门廊的表面全由砖砌成，但是砖块的摆放位置和方向不一。门廊的立面、地面还有顶部天花板都由砖作为饰面。另外，门廊由一个巨大的花岗岩柱子支撑。

New Crematorium at the Woodland Cemetery

The new crematorium is located in an undulating terrain in a wild wood section of the Woodland Cemetery, Stockholm. Surrounding the building is an area of massive, century old, pines. The building is a compact brick structure about 150 meters away from the major chapel complex by Eric Gunnar Asplund of 1940. The motto of the project in the anonymous international competition in 2009 was "A stone in the Forest". The arrangement of the plan and the characters of the interiors have been major concerns in the design of the new crematorium. The compact figure of the plan gives the staff overview as well as making a limited encroachment in the precious forest. Exposed white concrete has been used for the structure of the building. The aim has been to achieve the robust as well as to give a sense of clemency in the interiors. The concrete is from all-white Danish white cement and with Dolomite cross as ballast. No treatment has been done to the surfaces when the formwork has been dismantled. Thus walls

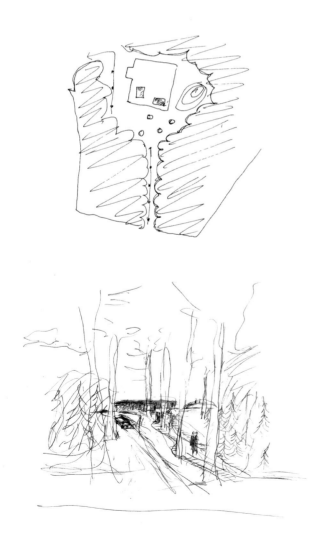

1. 信仰、希望与圣十字礼拜堂 2. 林中教堂
3. 游客中心，原来是服务建筑 4. 复活礼拜堂 5. 新火葬场
1. Chapels of Faith, Hope and the Holy Cross
2. Woodland Chapel 3. Visitor Center, previously service building
4. Chapel of Resurrenction 5. The New Crematorium

and ceilings give a subtle but palpable sign of the construction process.

For acoustics perforated bricks are used in some interiors. Being white glazed they reflect and accentuate the light from the openings and slits in the roof. One of the public spaces in the building is a Ceremony Room where mourners can have a ceremony by a coffin or urn. Inside the building block is an atrium open to the sky where staff can get together at breaks without interferring with mourners. Brick has been chosen for facades and roof to bring the small scale to the surfaces as well as for how it relates to the trunks of the surrounding pines. Mourners and visitors reach the building on a path of large granite slabs that are laid out between the pines. At the public entrance there is a generous brick canopy under which mourners may gather or rest in close proximity of the natural woodland. At the canopy all the surfaces are of brick laid on different edges. They are on the facades, on the ground and cast into the ceiling of the canopy. Adding to the surfaces of brick is one massive load bearing granite column.

项目名称：New Crematorium at the Woodland Cemetery
地点：Sockenvägen 492, Stockholm
建筑师：Johan Celsing Arkitektkontor / 建筑师负责人：Johan Celsing
项目团队：Stefan Andersson, Göran Marklund, Elisabet Bernsveden, Sven Etzler, Eyvind Bergström, Ibb Berglund, Tommy Carlsson, Kristina Dalberg, Marcus Eliasson, Milo Lavén, Sabina Liew, Thomas Marcks, Anna Ryf, Carl Toråker, Carl Wärn
结构工程师：Tyréns AB / 暖通空调：Anders Dahlbeck VVS Konsult AB
电气工程师：Sonny Svenson Konsult AB / 声学设计：Akustikon AB
景观建筑师：Rita Illien, Müller Illien Landschaftsarchitekten GmbH
客户：The Stockholms Cemetery Committee
有效楼层面积：3,000m² / 设计时间：2009 / 竣工时间：2013
摄影师：©Ioana Marinescu (courtesy of the architect) (except as noted)

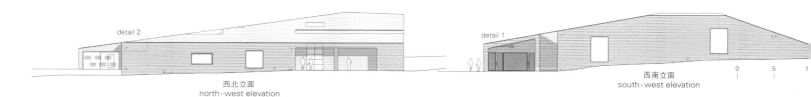

西北立面
north-west elevation

西南立面
south-west elevation

东南立面
south-east elevation

东北立面
north-east elevation

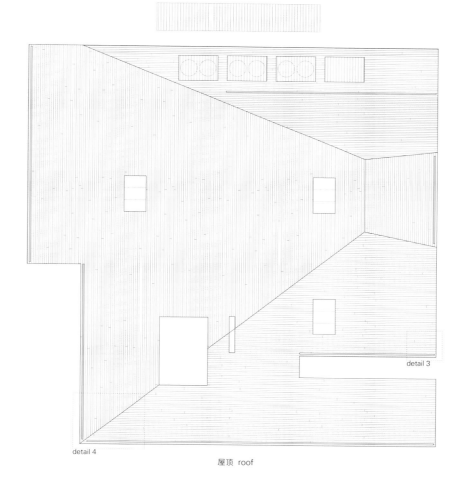

屋顶 roof

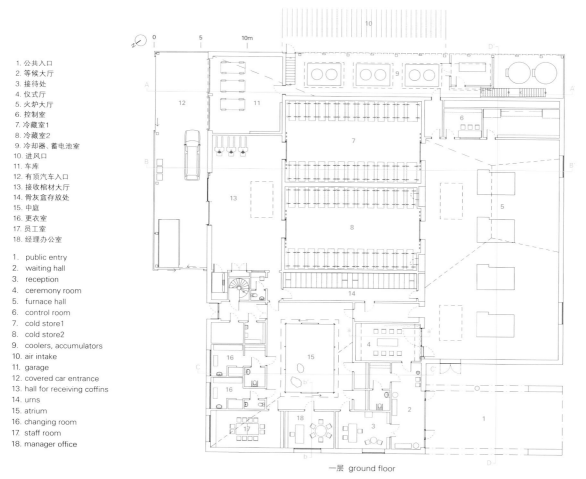

1. 公共入口
2. 等候大厅
3. 接待处
4. 仪式厅
5. 火炉大厅
6. 控制室
7. 冷藏室1
8. 冷藏室2
9. 冷却器、蓄电池室
10. 进风口
11. 车库
12. 有顶汽车入口
13. 接收棺材大厅
14. 骨灰盒存放处
15. 中庭
16. 更衣室
17. 员工室
18. 经理办公室

1. public entry
2. waiting hall
3. reception
4. ceremony room
5. furnace hall
6. control room
7. cold store1
8. cold store2
9. coolers, accumulators
10. air intake
11. garage
12. covered car entrance
13. hall for receiving coffins
14. urns
15. atrium
16. changing room
17. staff room
18. manager office

一层 ground floor

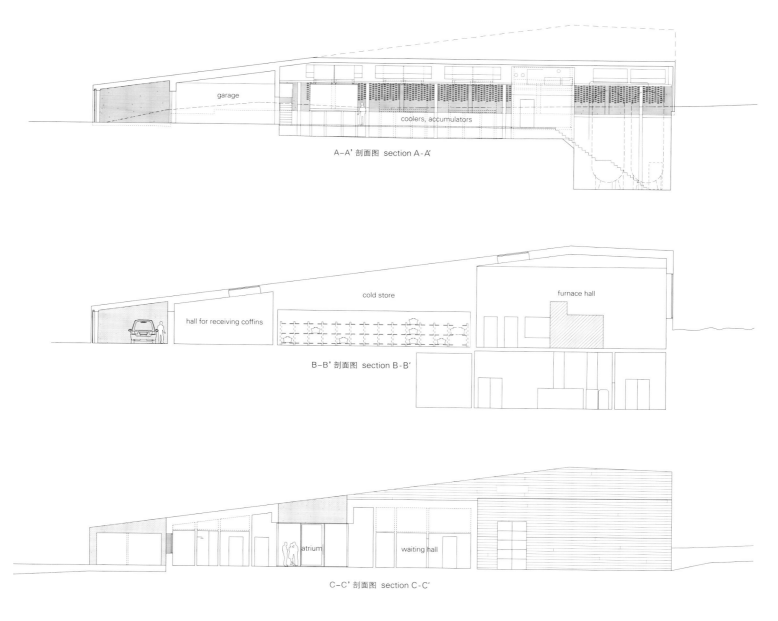

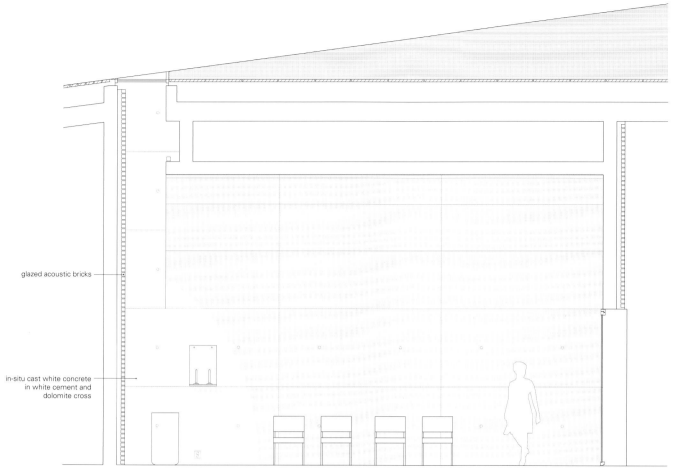

a-a' 剖面详图——仪式厅
detail section a-a'_ceremony room

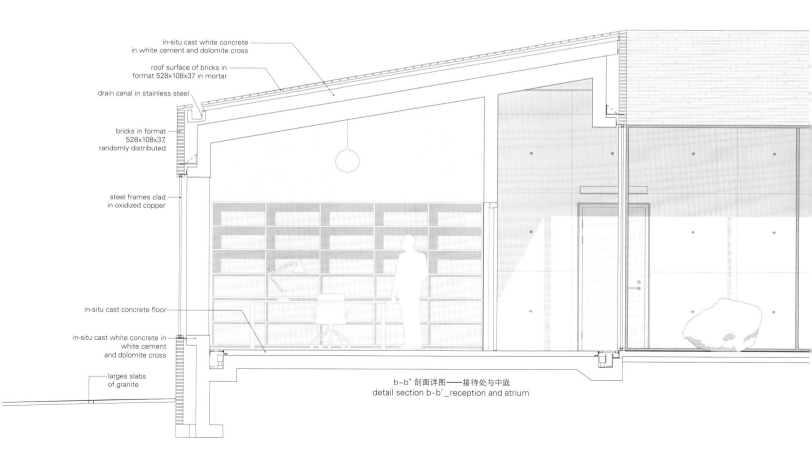

b-b' 剖面详图——接待处与中庭
detail section b-b'_reception and atrium

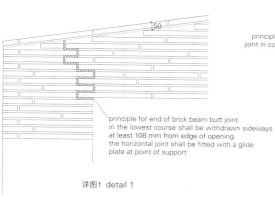

corner detail on roof, surfaces clad in bricks.
built-in dewatering gutters.

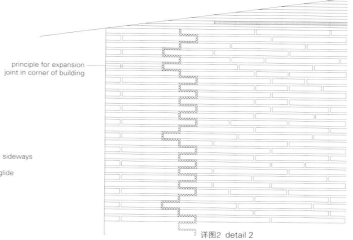

principle for expansion joint in corner of building

principle for end of brick beam butt joint in the lowest course shall be withdrawn sideways at least 108 mm from edge of opening. the horizontal joint shall be fitted with a glide plate at point of support

详图1 detail 1

详图2 detail 2

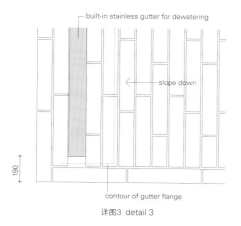

built-in stainless gutter for dewatering

slope down

contour of gutter flange

详图3 detail 3

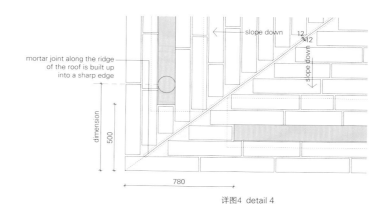

slope down

mortar joint along the ridge of the roof is built up into a sharp edge

slope down

dimension 500

780

详图4 detail 4

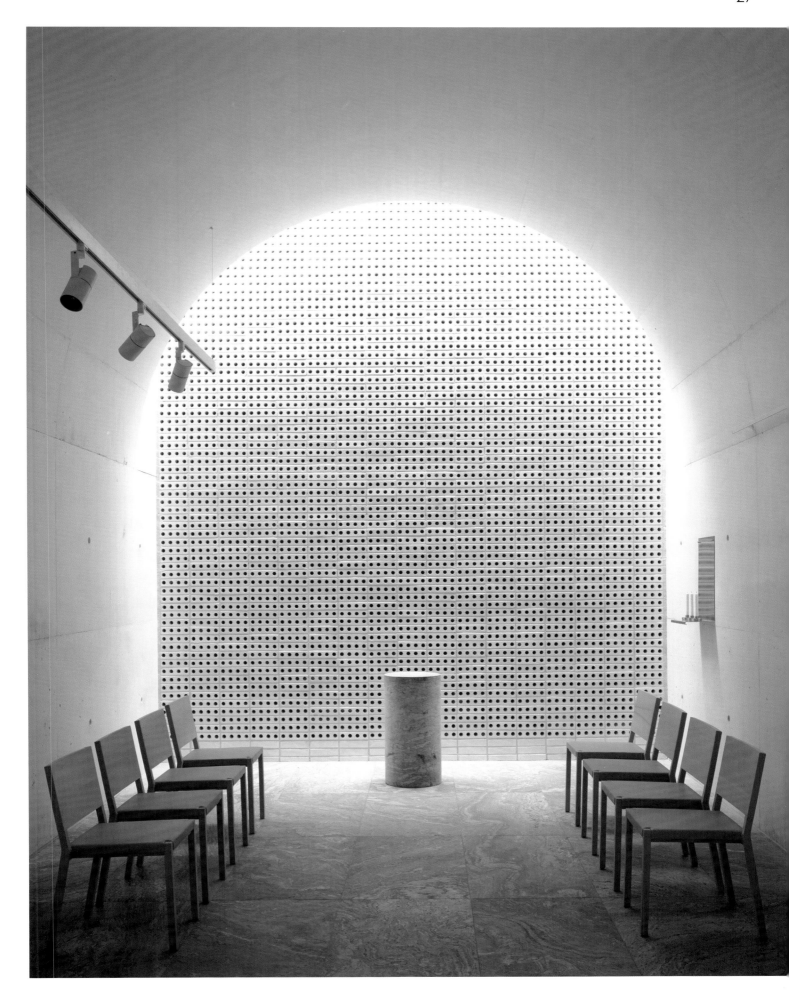

建筑情感：从宗教到世俗 Architecture of Sentiment – from the sacred to the human

灵斯泰兹公共火葬场
Henning Larsen Architects

丹麦灵斯泰兹火葬场的设计既考虑到了对逝者的尊重又考虑到了火化的程序。在大多数火葬场中，小礼拜堂区域是主要的焦点，而火葬的地方仅仅被放置在只有工作人员才能进入的功能区。而这个公共火葬场恰恰相反，如果愿意，逝者的家人和亲人可以跟着灵柩目睹整个火化过程。

棺材最先从灵车中被抬出来，然后被搬到一个安静平和的房间里，人们可以在此向逝者做最后的告别。在这个安静的房间里，逝者的亲友可以选择透过窗户望着棺材进入火炉室火化。这个12m高的火炉室内洒满阳光；面向树林的东山墙和西山墙上安装了高高的窗户，使室内的砖墙和花岗岩地板沐浴在阳光中。

火葬场的设计也考虑到了在这工作的员工。新的火炉室有着良好的室内小气候和工作条件，同时，也为逝者的亲人提供了肃穆的氛围。日夜更替，四季变化，这儿的人都能够享受到阳光。温暖的阳光照射到墙上，在起伏的天花板上流动。

该建筑的外观长而低矮，而高高的火炉室位于该建筑物的中间。火炉室嵌入地基，使建筑更加贴近人的尺度，操作起来更加方便，从而更加人性化。该火葬场占地50 000m²，毗邻丹麦历史性建筑物Kærup Gods，一个像公园一样的地方。

灵斯泰兹火葬场的建设满足了烟气净化新标准，并取代了该地区其他八个旧火葬场。设计重点强调灵活性，同时也考虑到功能和非宗教性空间，旨在兼容各种类型的纪念聚会活动。将来，周边的地方将建成墓地，这样，逝者的亲友们可以就近将火葬场的骨灰盒拿到墓地埋葬。

Communal Crematorium in Ringsted

The crematorium in Ringsted, Denmark is designed with respect and consideration for the cremation process. In most crematoria, the chapel area is the main focus, while the cremation itself is confined to a back-of-house, functional area. Instead, at this communal crematorium, family and loved ones may follow the coffin all the way through the cremation process.

The coffin is first carried from the hearse to a quiet and peaceful room for last goodbyes. From this quiet room, friends and relatives of the deceased have the option to follow the coffin and observe the cremation through a window into the furnace room. The 12-meter-tall furnace room is flooded with

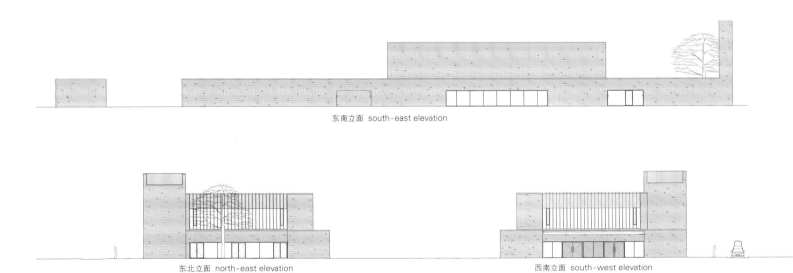

东南立面 south-east elevation

东北立面 north-east elevation

西南立面 south-west elevation

西北立面 north-west elevation

daylight; brick walls and granite floors are lit throughout the day by windows placed high on the east and west gables, which face the woods.

The crematorium is designed with consideration for the people who work there. The new furnace room creates good indoor climate and working conditions while also offering a worthy atmosphere for the relatives of the deceased. Users of the space are able to enjoy the light as it changes throughout the day and the year. The warm light reflects on the walls and flows in along an undulating ceiling.

From the outside the building volume is long and low, with the tall furnace room in the middle of the complex. The furnace room is inscribed into the base, bringing building to an approachable and human scale. Situated on a 50.000 square meter site, the crematorium is adjacent to the park-like grounds of an historic Danish estate, Kærup Gods.

The Ringsted Crematorium is built to meet new standards for flue gas purification and replaces eight former crematoria in the region. The design places an emphasis on flexibility, both with regards to function and to non-religious spaces, in order to sensitively accommodate all types of memorial gatherings. In the future, the surrounding site will function as a cemetery, and relatives will be able to pick up the urn from the crematorium and bury it at the cemetery.

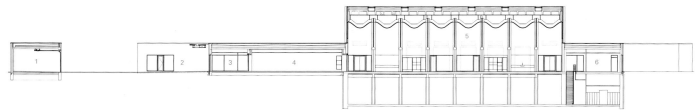

1. 车库 2. 汽车入口 3. 接收棺材大厅 4. 冷藏室 5. 火炉室 6. 办公室
1. garage 2. car entrance 3. hall for receiving coffins 4. cold store 5. furnace hall 6. office
A-A' 剖面图 section A-A'

项目名称：Communal Crematorium in Ringsted
地点：Ringsted, Denmark
建筑师：Henning Larsen Architects
结构工程师：Elindco
咨询工程师：Damgaard Consulting
景观建筑师：Birgitte Fink
客户：Communal crematorium Zealand and Lolland-Falster I/S
有效楼层面积：2,700m²
施工时间：2012—2013
摄影师：©Anders Sune Berg (courtesy of the architect)

1. 仪式厅 2. 火炉室
1. ceremony room 2. furnace hall
B-B' 剖面图 section B-B'

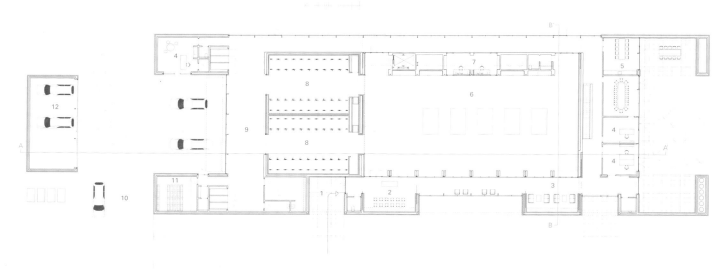

1. 公共入口 2. 等候室 3. 仪式厅 4. 办公室 5. 员工室 6. 火炉室 7. 控制室 8. 冷藏室
9. 接收棺材大厅 10. 汽车入口 11. 骨灰盒存放处 12. 车库
1. public entry 2. waiting room 3. ceremony room 4. office 5. staff room 6. furnace hall
7. control room 8. cold store 9. hall for receiving coffins 10. car entrance 11. urns 12. garage
一层 first floor

东立面 east elevation

亚眠火葬场

PLAN01

很少有包含如此丰富情感和表达如此丰富人性与文化内涵的建筑项目。然而，这样的设计也会遇到种种困难和风险，但是战胜这些困难和风险之后，生命却能够展现无穷力量。

在我们经历人生中倍感恐惧的时刻——失去所爱的人，最重要的是，在逝者的尸体被火化这一最真实而又极端的时刻，这种设计能够创造出一种庄严肃穆、平静安详的环境氛围，这正是其成功之处。

该建筑的设计基于大量的圆形图像的使用，因为圆形代表一些清晰可辨的主题，比如：逝者的中心地位；象征宇宙和永恒；世俗精神；脱离凡世生活；旅途的流动性和多样性；整个设计与景观融为一体，并且很巧妙地隐藏技术设备，尤其是烟囱。

建筑师对场地的特定诠释营造了一个中空空间，与环境共鸣，整体布局远近结合、规模大小和质地纹理不一，赋予该建筑高度灵活性。设计动静结合，定义了一个敏感而又人性化的灵活空间，宁静而祥和，轻快而明亮；自然植被与建筑物之间非常和谐，体现了场地和建筑项目之间的相互依存。

圆形的墙体和柔和的景观旨在与我们日常生活中常见的"直角世界"形成鲜明对比。

沙色的混凝土墙上，落地窗户镶嵌在金色的窗框内，面朝花园。花园中有两条方向相反的小路通往建筑物。这种布局是为了给参加不同葬礼的客人提供不同的路线。另外，还有一条路通往火葬场，但被隐藏在了这个地块后部的位置。相邻的还有两个停车场，但都隐藏在一排排植被的后面，从火葬场看不到。

整座建筑物由许多分布在一片空地上的圆柱形空间组成。围绕大厅和天井有两个仪式厅，可以举行不同规模的仪式。两个仪式厅都可

项目名称：Crematorium in Amiens / 地点：Avenue de Grâce – 80,000 Amiens, France
建筑师：Atelier Phileas and Ignacio Prego for PLAN01 (workshop of 5 agencies: Phileas, Atelier du Pont, Ignacio Prego Architectures, Jean Bocabeille Architecte and Koz)
项目经理：Clément Keufer / 生态设计：PLAN02 / 设备：Gem Matthews / 景观：Sempervirens / 技术设计：Grontmij Sechaud Bossuyt
总承包商：Léon Grosse / 客户：Amiens Metropole / 面积：1,285m² / 造价：EUR 4,810,000 (extra taxes) / 竞赛时间：2012.3 / 交付时间：2015.1
摄影师：©Takuji Shimmura (courtesy of the architect) (except as noted)

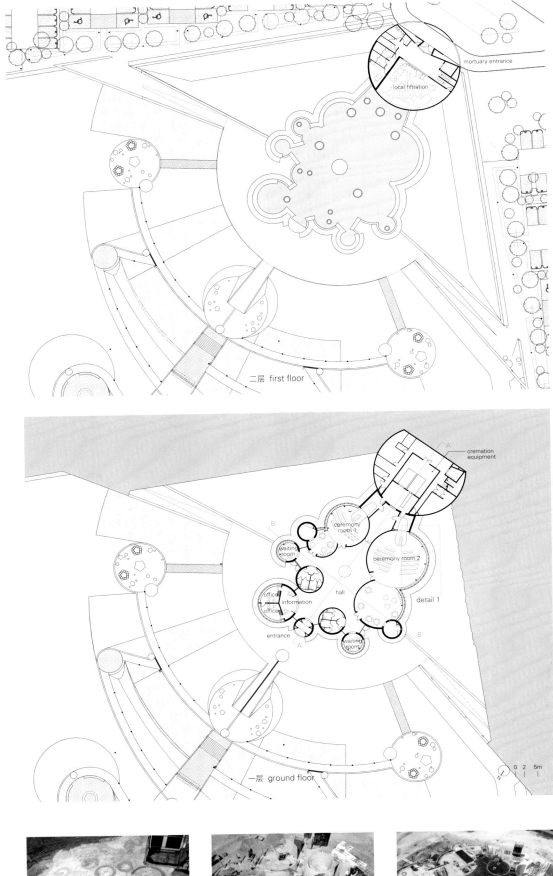

二层 first floor

一层 ground floor

以通到观看厅和提取骨灰的房间。作为仪式厅的直接延伸，会议室的布局使其自然成了仪式厅的延伸部分。该建筑非常模块化，用特定的布帘就能简单而又快捷地划分出不同的空间。

主中央大厅是仪式的焦点。主中央大厅设计包含大厅本身设计和通道设计：这里没有走廊，只有流动的空间，这里洒满通过天花板或者房间之间的窗户照进室内的光线。

所有空间的处理都绝对现代、温馨而简洁。其主要特点是明亮、金光闪闪，还有勾画出光影效果的雕塑般的混凝土。连绵不断的地面，白色的布帘，明净的玻璃，白色和金色相间的底面使光照间接而柔和。

Crematorium in Amiens

Few projects involve so much emotion and express such human and cultural richness. However, there are also as many obstacles and risks which, when overcome, allow life to appear in all its power.

The success of this approach lies in its ability to create the environment of a dignified and serene experience during one of the most dreaded moments of our existence – the loss of a loved one – and, what is more, during one of the most tangible and extreme moments, when the body of the deceased person is cremated.

Our architecture is based on this strong image of the circle, which is representative of clearly identifiable themes, such as, the central position of the deceased person, the circle as a universal and timeless symbol, secular spirituality, removal of references to daily life, the fluidity and diversity of the journey, blending into the landscape and hiding the technical elements, particularly the chimney.

Our specific interpretation of the site has given rise to a hollow space which resonates with its environment, an arrangement which allows great flexibility due to its combinations of near and far and its range of scales and textures. A fluid and calm dynamic, a flexible definition of a sensitive and considerate space. A serene and light response, interde-

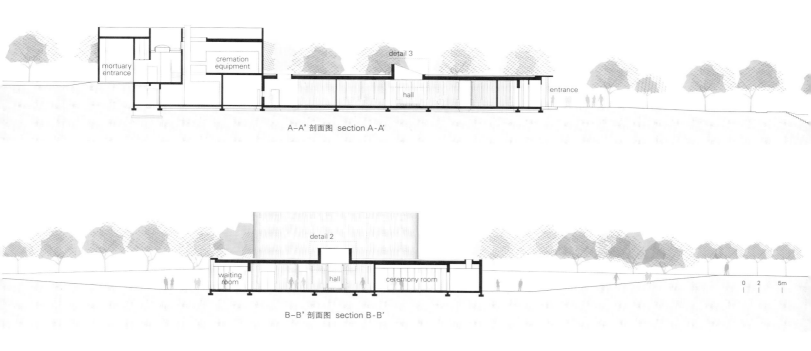

A-A' 剖面图 section A-A'

B-B' 剖面图 section B-B'

pendence between the site and the project, is demonstrated through the harmonious relationship between vegetation and the building.

The rounded walls and soft landscaping are intended as a distinct departure from the "world of right angles" encountered in everyday spaces.

Full-height windows sit within golden frames in the sandy-toned concrete walls. They face out towards the gardens, where two pathways approach the building from opposite directions. This layout provides separate routes for guests of different funerals, while a third access route leading to the cremation area is concealed at the back of the plot. A pair of car parks also adjoin the site, but are hidden from view by screens of planting.

The building is a collection of cylindrical spaces laid out in a clearing. The two ceremony rooms surround a hall and a patio which are provided for different sized ceremonies. Each group can access the viewing room and the room for handing out the ashes through the ceremony rooms. As a direct extension of the ceremony rooms, the meeting rooms are set out so that they can naturally become extensions of the ceremony spaces. The building is very modular and is partitioned simply and swiftly with specific fabric drapes.

The main central hall is the focus of the ceremony, including both the hall itself and the movement: there are no hallways but fluid spaces which are lit with natural light through the ceiling or the windows between the rooms.

The spaces are handled in a resolutely contemporary, warm yet simple manner. The dominant characteristics are light and golden aspects with sculpted concrete which draws out the shadows and the light: continuous flooring, white drapes, clear glass, a white and gold underside which provides indirect and softened lighting. PLAN01

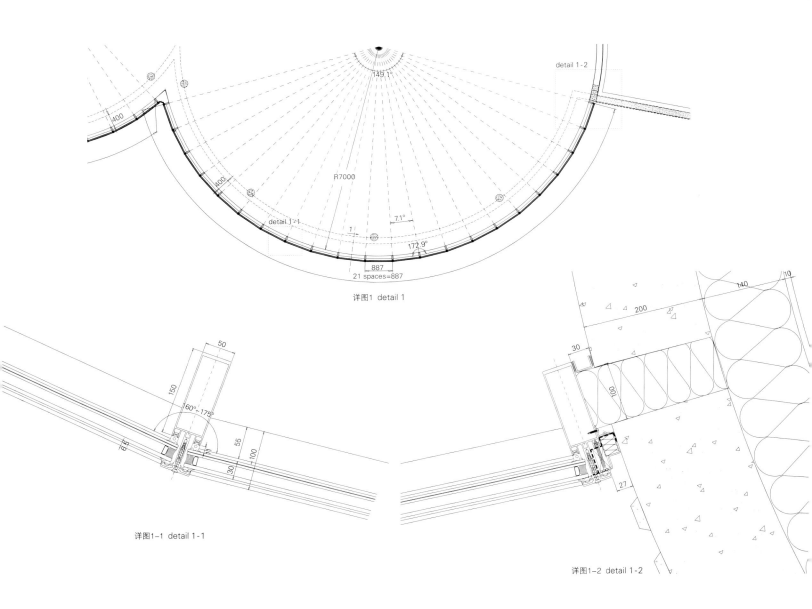

详图1 detail 1

详图1-1 detail 1-1

详图1-2 detail 1-2

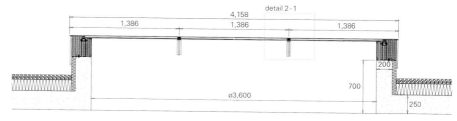

详图2 detail 2

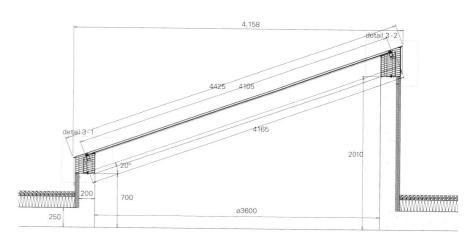

详图3 detail 3

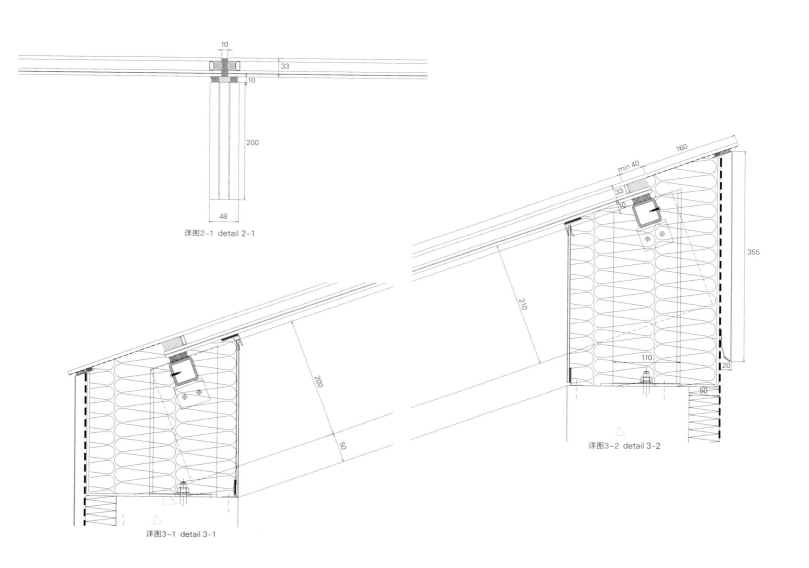

详图2-1 detail 2-1

详图3-1 detail 3-1

详图3-2 detail 3-2

Mahaprasthanam印度教火葬场和公墓

DA Studios

西南立面——入口建筑与行政办公区
south-west elevation _ entrance pavilion and administration

受凤凰基金会和市政府的委托,本项目对一个疏于管理的火葬场进行现代化改造。现存的火葬场/墓地占地1.5ha,本项目就是对其进行现代化升级改造,以适应印度教火葬仪式的文化和环境的现代化需求。追求清晰、时间限制和不可抗拒的命令——体现现代性,这一切使设计师只能选择坟墓间四处零散而荒芜的地块加以改造,从而使整体设计形成分散状效果。然而,这正是设计的优势所在,按照印度教仪式功能要求布局,同时体现现代化策略要求。

在印度教哲学体系中,一个人只有经历了被称为Shodasha Samskara的16个阶段之后,他的人生目标才算得以实现,而最后一个阶段就是被称为Antyesti的葬礼仪式。葬礼仪式由五个主要部分组成:尸体准备、火化过程、哀悼仪式、亲人净化和回忆/哀悼期。

人们认为在对真理和完美的永久追求中,行动、责任或者结果是不相关的。

该项目是对逝者的敬畏、怀念和接纳的真实而具体的延伸。设计从入口处的仪式化列队行进小路开始,小路两侧是两堵庄严的稍微倾斜的墙,用来表达对逝者的尊重和敬畏。这些预制混凝土墙上刻着神圣的Gta经文,这些经文暗示一股神奇的力量,能使逝者看到将来如诗如画的生活和留之身后的情感。墙拔地而起,最低处4m高,引导人们的视线望向火葬用的柴堆。过了这两堵墙的左侧,一个内凹的空间里是

1. 主入口 2. 冷藏室 3. 车行道 4. 入口建筑 5. 电动化火葬场 6. 行政办公区 7. 火葬场庭院 8. 等候厅1 9. 葬礼室1——传统 10. 更衣室/淋浴间/卫生间 11. 等候厅2 12. 葬礼室2——传统 13. 食堂 14. 干景观 15. 景观 16. 公共火葬柴堆 17. 原有坟墓 18. 停车场
1. main entrance 2. cold room 3. drive way 4. entrance pavilion 5. electrical crematorium 6. administration 7. ceremonial yard 8. waiting hall 1 9. funeral pier 1_traditional 10. change rooms/showers/toilets 11. waiting hall 2 12. funeral pier 2_traditional 13. canteen 14. dry landscape 15. landscape 16. public funeral pyre 17. existing graves 18. parking

东北立面——入口建筑与行政办公区
north-east elevation_entrance pavilion and administration

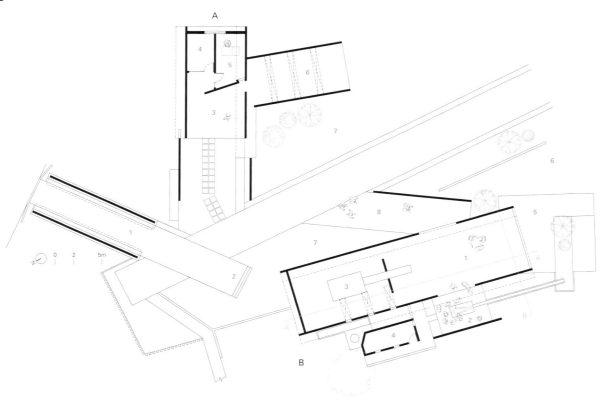

A. 入口建筑与行政办公区
 1. 入口建筑
 2. 画壁
 3. 办公楼
 4. 书店
 5. 管理人员办公室
 6. 骨灰堂
 7. 景观

B. 电动化火葬场
 1. 电动化火葬场
 2. 葬礼仪式庭院
 3. 火炉
 4. 机械室
 5. 开放庭院
 6. 干景观
 7. 景观
 8. 小路

A. Entrance Pavilion & Administration
 1. entrance pavilion
 2. mural
 3. office block
 4. book store
 5. managers room
 6. locker room
 7. landscape

B. Electrical Crematorium
 1. electrical crematorium
 2. ritual yard
 3. furnace
 4. machine room
 5. open yard
 6. dry landscape
 7. landscape
 8. pathway

行政办公区域所在，由一栋办公楼、一个书店和一个骨灰堂组成。骨灰堂大约有100个用来纪念和存放骨灰的骨灰柜。这个地方从原来杂乱无序的状态，被重新注入活力，将印度教葬礼仪式的各个步骤重新协调地组合在一起。

每一个步骤都在提醒着我们并没有失去所爱的人，而是我们所爱的人的灵魂从此以后以不同的形式继续存在着。因此，对原有的火葬场内的空间加以设计导致建筑布局四处分散开来。预制结构的空窗设计突出了仪式和葬礼是一个清晰的流线型过程。比生命还要庄严的建筑体量似乎能为失去亲人的人们带来安慰，随着仪式的进行，人们变得更加平静，更容易接受现实。

此项目场地的动态性要求这些巨大的纪念物能反映出这一动态特性。穿过入口处两墙之间的小路，一座正在倒下的混凝土建筑进入人们的视野。这一拥有进口技术的电动化火葬柴堆建筑就属于这一类。正在倒下的建筑形式并不是偶然，而是专门建造来向逝者致敬的。其空窗设计使人联想到了沿山而建的佛教僧侣的佛教寺庙殿堂以及佛塔。

沿着这条仪式化列队行进小路继续向前走，你会发现这个地方的其他特点逐一展露出来：宁静祥和的绿化景观，令人浮想联翩的经文，坟墓里的记忆。因其功能（哀悼逝者）非常复杂，所以自然火葬柴堆这部分的设计也更为复杂。这些建筑是这样设计的：人们在等候厅里安静地鞠躬哀悼，而火葬柴堆建筑呈开放状，方便人们向逝者告别。这两座建筑是分开的，用以划分参加葬礼的不同人群，这样也能让人们明确区分在这些空间里所举行的不同仪式程序。等候厅可容纳近300人，通向火葬柴堆的小路只有逝者最亲近的家人才能走，而只有一家之长和主持火葬仪式的神父才能到火葬柴堆旁。

Mahaprasthanam Hindu Crematorium and Cemetery

The client Phoenix Foundation and the City Municipality undertook the task of accomplishing this state of the art crematorium from a neglected stage. As an already existing 1.5ha crematory/cemetery, the initiative was to modernize it so that it can accommodate the cultural and contextual modern needs of Hindu cremation rituals. A search for clarity, time constraints and impeccable order – being the modern outlook lead to picking up untouched land pockets amongst the graves, thereby leading to a scattering effect on the site. Yet, it is to an advantage as it supports the modernization in strategic laying along the functions for the Hindu rituals.

In Indian Philosophy, purpose of life is fulfilled when one goes through the 16 phases called Shodasha Samskara, and the final stage is the funeral ritual called Antyesti. That consists of five major stages: preparation of the body, cremation process, rites of mourning, purification for the family members, remembrance/mourning period.

It is a belief that in the eternal search for truth and perfection, the action, duty or consequences are irrelevant.

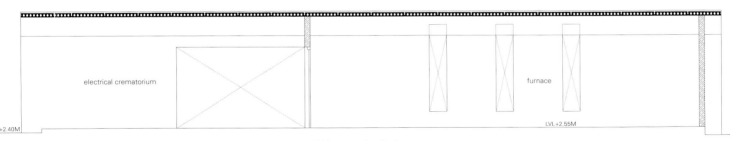

A-A' 剖面图 section A-A'

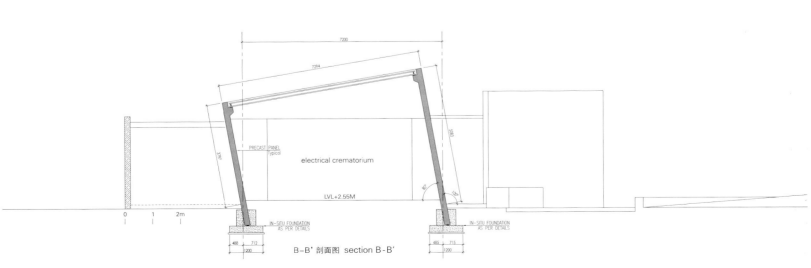

B-B' 剖面图 section B-B'

The project is a bodily extension of reverence, memory and of acceptance. It begins right at the entrance of the processional pathway with two stately walls inclined in a fashion of respect and revere of the dead. These precast concrete walls engraved with sacred Gta scriptures hint upon the strength to behold the picturesque scenery of life ahead and the emotions left behind. Raising above the ground to a height of 4 meters (on the shallow end) they direct one's eyes towards the cremation pyres. Towards the left of these lone walls, in a cove of space is the administrative block comprising an office building, a book shop and a locker room with nearly 100 lockers to pay respects and store ashes. From a state of entropy the site was rejuvenated and restructured for an orchestrated procession of the Hindu funeral ritual.

Every step is a reminder not of lost but of love as spirit of the loved lives on in many forms hence forth. Hence, designing spaces amongst existent crematory lead to scattered laying of built form. Cavumaedium of precast structures highlights the ritual and funeral procession as a linear and clear proceeding. The stately presence of a volume higher than life seemingly provides a comfort to the bereaved as more quite and acceptance comes with each step.

Dynamic nature of the site dictate monolithic monuments reflecting its nature. After crossing the processional path under the entrance walls, a fallen concrete form catches the eye. The Electrical Pyre with its imported technology is one of its kind. The afore mentioned fallen form is not accidental but an intentionality of salutation to the dead. This form with its cavumaedium reminds one of the Chaitya Halls of the Buddist monks with their Stupa located along the line of the mountains.

As one walks further along the processional pathway several other aspects of the site unfold themselves; serene landscaping, reminding scriptures and memories in graves. the natural funeral pyre is much elaborated due to its complex functionality, lamenting the lost, the structures are shaped in such a way: Waiting halls silently bow in honour, Pyres open up to liberate the lost in a farewell. These two structures are separated due to the distribution of crowds ergo giving lucidity to the rituals undertaken in these spaces. Waiting halls can accommodate nearly 300 people while the pathway leading to the pyre is meant for the closest of the family members and the pyre is meant for the head of the family along with the priest who overlooks the ceremony.

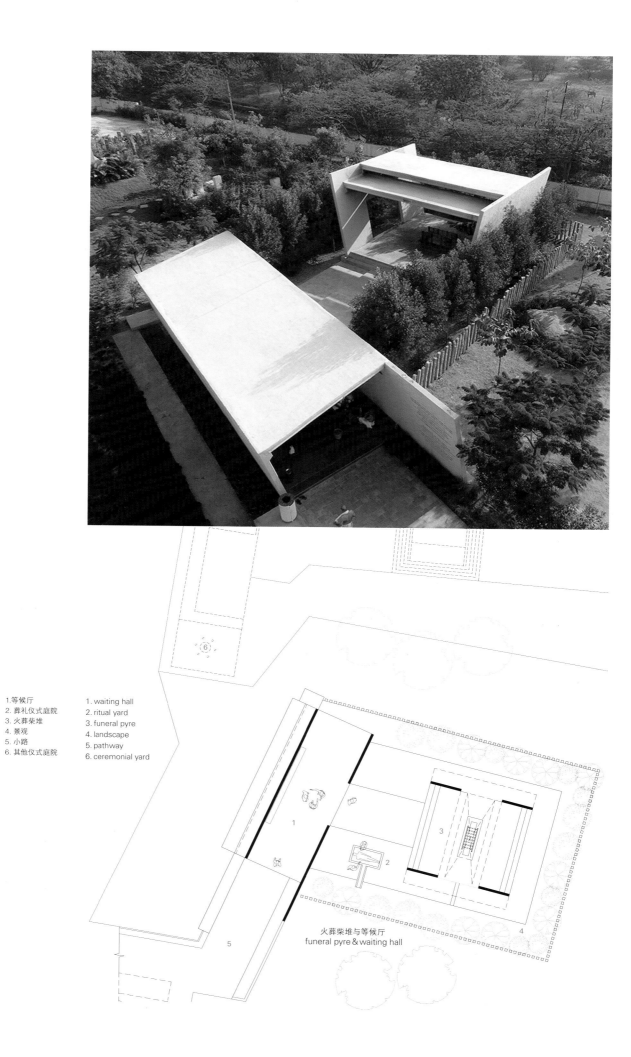

1. 等候厅
2. 葬礼仪式庭院
3. 火葬柴堆
4. 景观
5. 小路
6. 其他仪式庭院

1. waiting hall
2. ritual yard
3. funeral pyre
4. landscape
5. pathway
6. ceremonial yard

火葬柴堆与等候厅
funeral pyre & waiting hall

项目名称：Mahaprasthanam Hindu Crematorium and Cemetery / 地点：Hyderabad, Telangana, India / 建筑师：DA Studios / 项目团队：Chaitanya, kasi, Pradeepthi / 承包商：Preca and Phoenix Foundation, Hyderabad / 客户：Phoenix Foundation in collaboration with Greater Hyderabad Municipal Corporation, built under corporate social responsibility inititative / 项目类型：A Hindu crematorium, traditional and electric / 技术：precast concrete panels, precast hollow core slabs / 材料：pre cast concrete panels, gravel finished masonry walls, terracota tiled pathway / 建筑面积：3.7 acres / 竣工时间：2015.4 / 摄影师：©Sameer Chawda (courtesy of the architect)

公墓锯齿状混凝土顶棚
Ron Shenkin Studio

建筑情感：从宗教到世俗 Architecture of Sentiment – from the sacred to the human

该建筑是哀悼者聚集以及葬礼前和葬礼期间诵读哀悼词的地方。其主要特点是:它是一个凉亭结构或者说是一个开放的侧面有遮挡的棚子,紧挨公墓。

该建筑有两个入口,其中较小的一个供逝者家人使用,而较大的一个供其他哀悼者使用。

这样的入口设计使人们很容易进入这座宽敞的建筑。建筑的前部朝北敞开。建筑的位置也可以满足那些更愿意保持一定距离却又感觉离哀悼现场很近的哀悼者们的需求,同时也在建筑外提供了一片阴凉之地。有两个出口通向墓地,其中一条很容易走,另一条则需要通过一段台阶。该建筑西面有一堵坚固的石墙,可以阻止下午西晒的阳光,只有一部分阳光可以穿过狭长的窗户照进来。

该建筑所处的地方被一片果园包围着,果园最初规划是建新的住宅和商业楼。在设计的过程中,建筑师决定要充分考虑周围环境因素。因此,决定在这为了城市发展而遭到破坏的农村地区建一座纪念性建筑,恢复由于城市建设而拆除的乡村环境。

顶棚由裸露的混凝土板构成,象征城市建设的扩张和延伸。混凝土板由树状的金属柱支撑,代表那些被砍伐的树木。场地内保留了一棵橡树。橡树上面没有混凝土天花板,这一部分是开放的,让树自由生长,象征树木和这些金属标志之间的对话。

建筑北侧有一条混凝土线,自西侧的地面开始,经过窗户,一直向上延伸至屋顶、天花板和墙面,最后又很自然地回到地面——就像人的一生一样,来自尘土,又回归尘土。建筑的材料和色调都是单色的,主要由灰色调构成,给人一种整洁、中性和悲伤之感。夜晚,建筑物内部灯光亮起,熠熠生辉。

该建筑共使用了300多块大小、形状不同的板材,通过电脑软件拼接在一起。这些板材被运到工地,由承包商团队组装完成。金属"树"结构也是如设计那样加工好后运到工地现场组装完成的。等所有的铁艺和结构设计得到工程师的核准认可后,才用剩余的托板进行墙体加工。墙体和屋顶的混凝土浇筑需要一天的时间。

Jagged Concrete Canopy at a Cemetery

The structure functions as a place of convergence of mourners and for the reading of eulogies prior to and during the burial. Its essential nature is that of a pavilion or open sided shelter and is located next to the cemetery.

The building has two entrances. The smaller of the two serves the family of the deceased and the second and larger one serves the rest of the mourners.

This entry allows easy access to the spacious building. One side of the front of the building remains open to the north. The positioning of the structure also allows those mourners who prefer to remain at a distance, a sense of closeness to the proceedings and a shaded area that is not inside the structure itself. There are two exits leading to the burial lots, one of which is easily accessible and the other is via a flight of stairs.

一层 ground floor

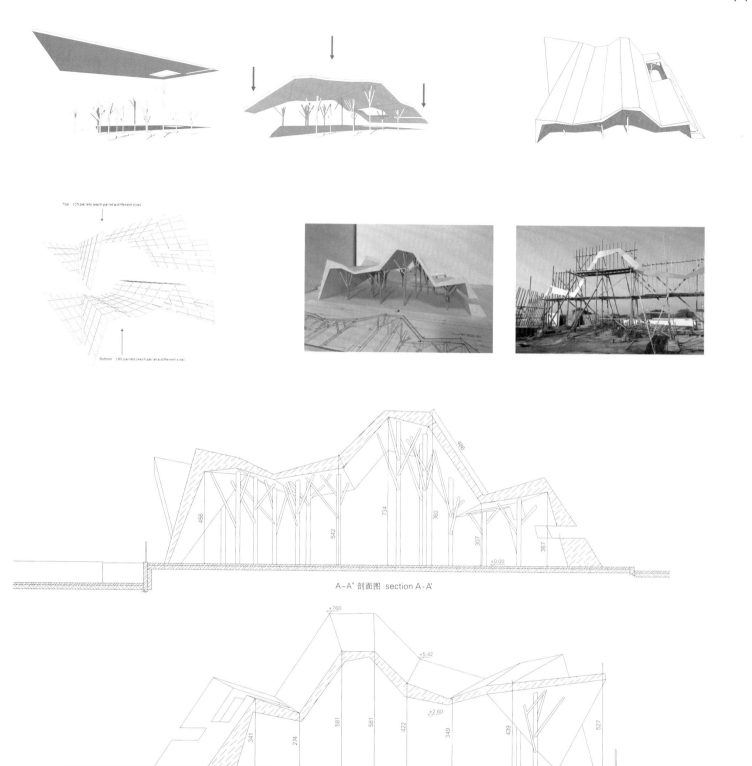

A-A' 剖面图 section A-A'

B-B' 剖面图 section B-B'

The structure has one solid stone wall to the west which serves to block any direct afternoon sunlight except what is allowed through via a long narrow window.

The area in which the structure is situated was surrounded by orchards which were originally raised to make space for new residential and commercial buildings. During the design process we decided to take the surroundings into consideration by erecting a monument to the rural surroundings that were demolished for urban repurposing.

The monument consists of an exposed concrete slab symbolizing the expansion of construction. The slab is stabilized by tree-shaped metal pillars denoting the trees that were cut down. One oak tree remains within the structure. The ceiling above the tree was left open to allow for the presence of the tree to create a dialogue between the living tree and the metal symbols.

On the northern side of the building there is a line of concrete, beginning in the ground on the west side and climbing up through the window. Ascending to the roof, ceilings and walls and making its final decent back to the ground – like a man who comes from dust and to dust returns. The building materials and color pallet are Monochromatic consisting mostly of shades of grays and providing a sense of cleanliness, neutrality and sadness. During the evening the building is lit from the inside and out.

The structure is designed using more than 300 panels of differing sizes and shapes and fitted together by computer software. The panels were brought to the construction site and assembled by a team of contractors. The metal "tree" structures were brought to the site and assembled as per the design. All iron works and structural designs were approved by engineers after which the walls were constructed by the remaining pallets. The concrete for the walls and roof was poured in one day. Ron Shenkin

项目名称：Jagged Concrete Canopy at a Cemetery
地点：Pardesiya, Israel
建筑师：Ron Shenkin studio for architecture & design
承包商：A.D. Haled
建筑面积：3,219m²
施工时间：2014—2015
摄影师：©Shai Epstein (courtesy of the architect)

建築情感：从宗教到世俗 'Architecture of Sentiment – from the sacred to the human

狭山森林小教堂
Hiroshi Nakamura & NAP

狭山湖畔墓地对各种各样的宗教和教派都是开放的。它坐落在大自然的环境中，这里接近水源涵养林，而场地本身就处于一片茂密森林的前面。建筑师心中所设想的建筑是这样的：它能反映生命的方式，生命靠水滋养，受到森林的庇护，死后最终又回到这个地方。随即建筑师发现，在各种各样的宗教中，森林都是人们祈祷的对象，于是就设想一栋被树木所包围、向森林祈祷的建筑。

该场地是一个小三角形地块，邻近一个交通不是很繁忙的市政路段和一条几乎没有行人的无名街道。这片森林本身已经超凡脱俗。因此，建筑师决定设计一个投入森林怀抱的空间，并通过把建筑墙体向内倾斜来避开树枝和树叶，从而形成日本传统的Gassho结构形式（即人们合掌祈祷的形状，译者注）：由成对的倾斜的梁从各个方面上相互支撑而形成的三维空间。屋顶使用由工匠手工做成的波纹状纹理的铸铝砖覆盖。

地面稍稍向森林这侧倾斜1cm，引导人们面朝死者坟墓的方向弯腰祈祷。地面石板的样式和接缝都完全一样，向前延伸直至消失在森林深处，这样能让人们把精力专注在森林上，专心祈祷。当人们祈祷时，双手的指尖轻轻并拢，手下方形成一个小而温暖的空间。建筑的造型就是这个空间的放大。当人们祈祷时，建筑跟着人们一起祷告。

Sayama Forest Chapel

Sayama Lakeside Cemetery is open to various religions and denominations. It is located in a nature-rich environment adjacent to the water conservation forest, and the site itself is in front of a deep forest. I envisioned an architecture that reflects the way of life as it lives by the water conserved by the forest, and eventually returns to this place after death. Thereupon, I found the forest to be the subject of prayer that is mutual to various religions and conceptualized an architecture that prays to the forest while surrounded by trees.

The site is a small triangular plot of land that is adjacent to a

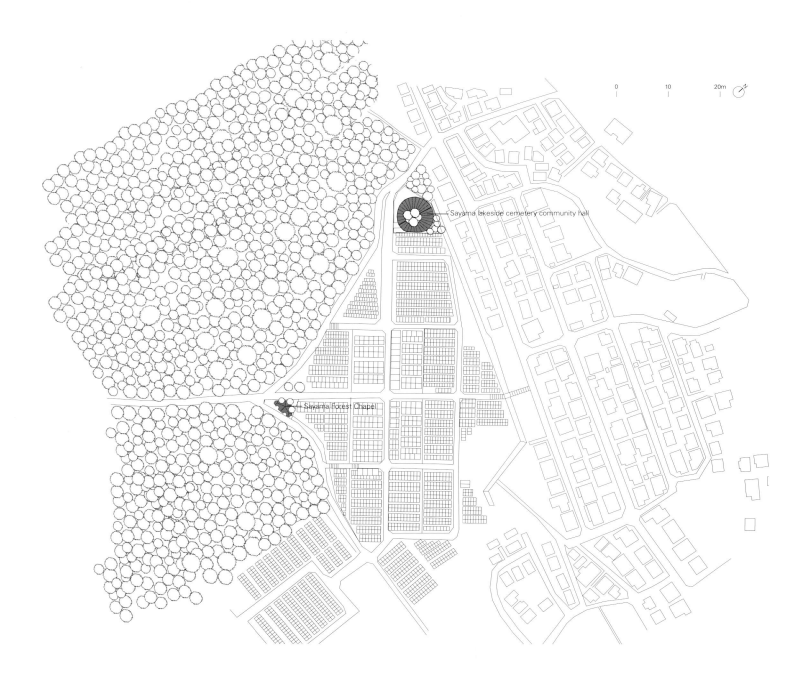

项目名称：Sayama Forest Chapel
地点：2002-4 Maekubomine, Kamiyamaguchi, Tokorozawa city, Saitama prefecture
建筑师：Hiroshi Nakamura & NAP
建筑与设计：Hiroshi Nakamura & NAP - Hiroshi Nakamura, Kohei Taniguchi, former NAP members - Atsushi Ikawa, Eisuke Hara, Keisuke Minato
主要用途：Chapel, Ossuary
客户：Boenfukyukai Foundation
结构工程师：Arup (Hitoshi Yonamine, Tetsuya Emura)
建筑设备：Arup (Kentaro Suga, Ayako Tanno, Makiko Arai, Celso Soriano Junior)
监理：Hiroshi Nakamura & NAP
施工：Shimizu Corporation
用地面积：148.22m² / 建筑面积：69.32m²
总楼面面积：110.49m²
地下层面积：41.17m²
地面层面积：69.32m²
建筑覆盖率：46.76%(allowed: 60%)
容积率：74.54%(allowed: 100%)
楼层：one above ground, one underground
结构：Steel-reinforced concrete and wood construction
设计时间：2011.4—2012.9
施工时间：2012.12—2014.1
摄影师：©Koji Fujii (courtesy of the architect)

municipal road with a low traffic and a non-legal street with almost no pedestrian. Therefore, I have decided to create a space that devotes to the forest that is transcendent in its existence, by tilting the wall inward to avoid the tree branches and leaves. It forms a traditional Japanese Gassho-style structure composed three-dimensionally as two leaning beams set against each other are developed in every direction. The roof is covered with cast-aluminum tiles with ripple-like textures each made by hands of craftsmen.

The floor inclined towards the forest by 1 centimeter guides people towards the departed and the forward bending posture for praying. The patterns and seams of the slate extend towards the vanishing point deep into the forest to help one concentrate the mind on the forest. When one prays, a small warm space is created within the hands as the fingers gently join. It seems as if that small space of prayer was taken out to form the architecture. As people pray, so does the architecture.

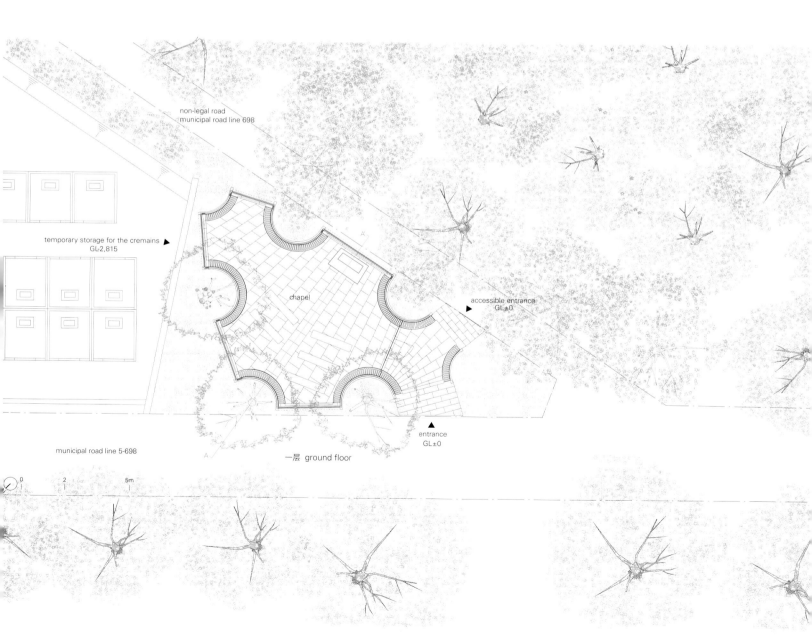

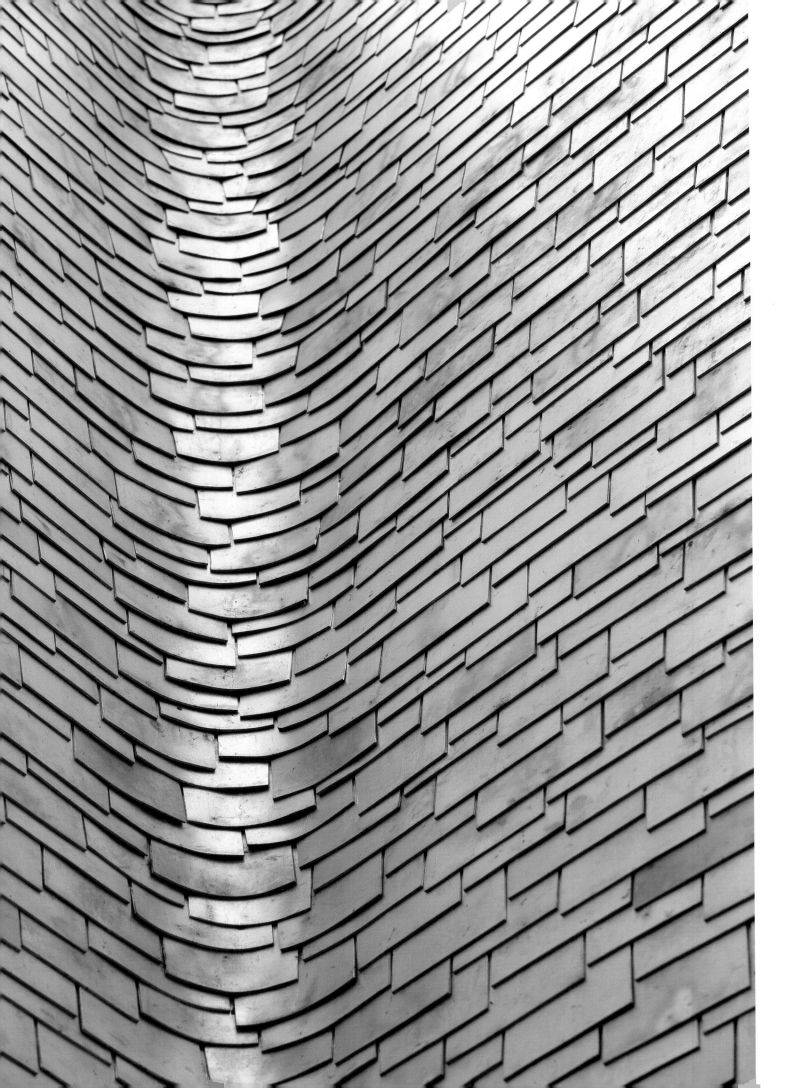

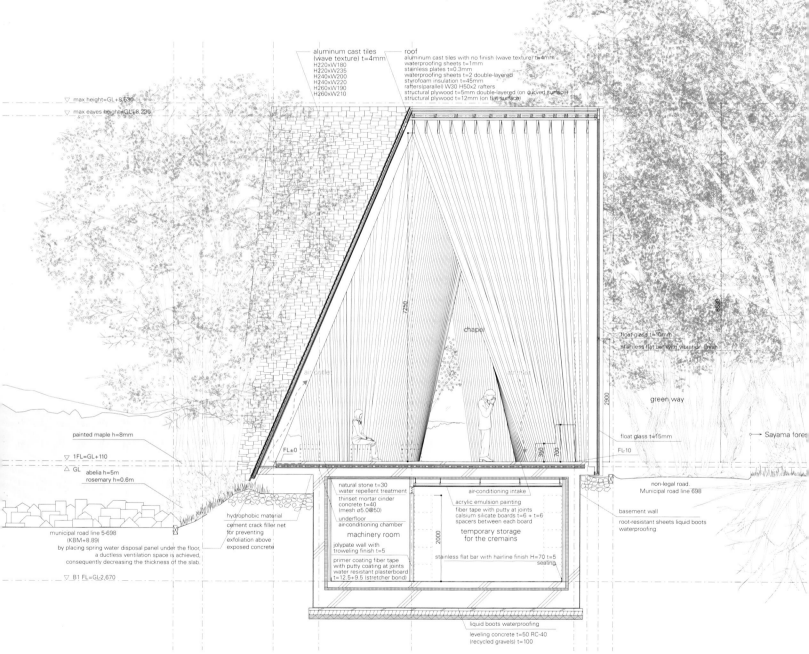

A-A' 剖面图 section A-A'

里韦萨尔特纪念馆
Rudy Ricciotti + Passelac & Roques Architects

除了天空，看不到任何外面的东西。然而，四处都有社交的阴影。三个天井分别构成了学习实验室，社交区域和办公区域，都提供了某种舒适感。该项目设计完全基于"接受"这一理念：接受这一区域(片区)、路线、变形异化的军事几何体，当然，也接受它的历史。里韦萨尔特纪念馆位于大地和天空之间，连接过去和记忆，恰恰就存在于现在，存在于生活本身。其形式上的暴力表明遗忘是不可能的。

The Rivesaltes Memorial

A witness to some of the twentieth century's darkest moments – the Spanish Civil War, World War II, the Algerian War of Independence – the Camp de Rivesaltes occupies a unique and important place in French history. A former military camp (Camp Joffre), a camp for Spanish refugees, the largest internment camp in Southern France in 1941 and 1942, an internment camp for German prisoners of war and collaborators, and the primary relocation center for Harkis and their families, its history is unique.

In order to tell this story, a memorial opened its doors on 16 October 2015. Built on the former block F of the camp, in the middle of the existing buildings, the memorial, measuring 4,000 sqm, provides an authoritative account of the history of the forced displacement and subjugation of populations. It is also a place where visitors may cultivate the memory of all those who once passed through its doors.

The memorial is silent and oppressive: it lies in the earth, squarely facing block F, with a calm and silent determination, a monolith of ochre-colored concrete, untouchable, angled towards the sky. At once buried in, and emerging from the earth, the memorial appears on the surface of the natural landscape as one enters the camp, and stretches to the east-

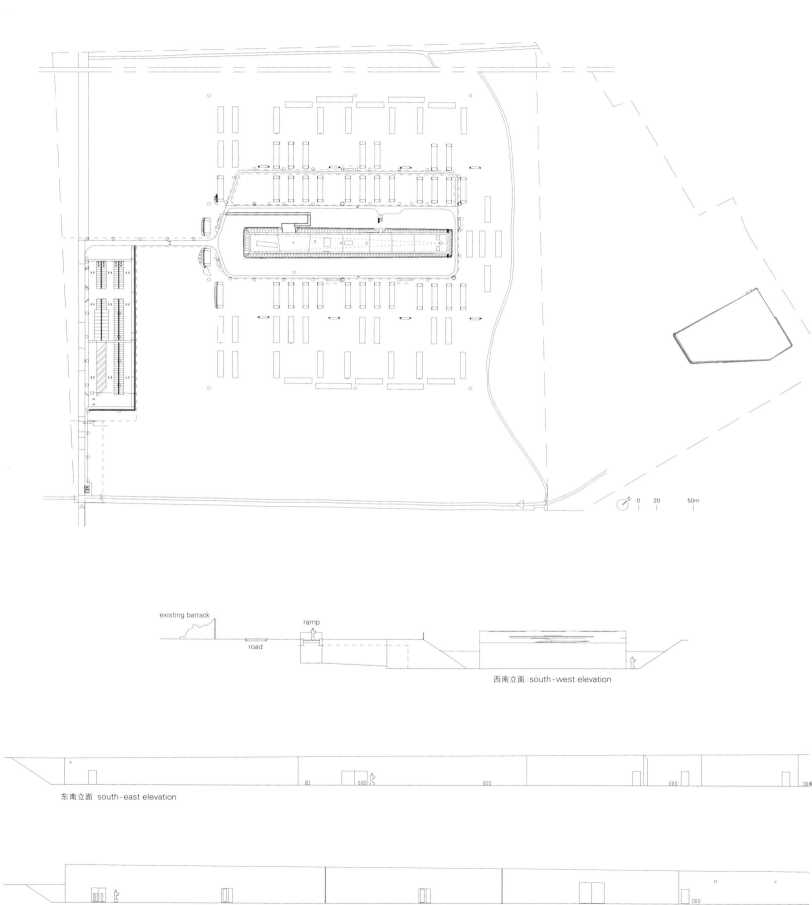

东北立面 north-east elevation

项目名称：The Rivesaltes Memorial / 地点：Rivesaltes, France / 建筑师：Rudy Ricciotti + Passelac & Roques Architects
项目管理：Regional Languedoc Roussillon, Roussillon Aménagement / 功能：permanent exhibition space, temporary exhibition space, 145-seater auditorium, research center, learning labs space, social area / 造价：EUR14.3M (H.T.) / 面积：3,590m² (1,500m² for exhibition) / 竞赛获奖时间：2005.8 / 施工时间：2013.1 - 2015.10
摄影师：©Kevin Dolmaire (courtesy of the architect)-p.96~97, p.98~99, p.101, p.102~103, p.104~105, p.110~115,
©M. Hedelin / Region Languedoc-Roussillon (courtesy of the architect)-p.107 / ©Olivier Amsellem (courtesy of the architect)-p.106, p.108~109

ern extremity of the former meeting place, to a height that is equal to the height of the roofs of the existing buildings. To the west of the memorial, some of the buildings have been rebuilt, recreating the serial and alienating spatiality of the camp. Here, there is an absence of vegetation, resulting in a flat, arid landscape, unmarked by shadows, and buffeted by the wind. From the carpark, situated at the outer southwest corner of the block, the visitor can enjoy panoramic views of the camp.

The memorial is reached by a pathway that starts from the carpark, in line with the entrance to the building. This pathway leads to either the entrance of the camp or to an exterior pathway or route, with views of the nearby Corbières and Pyrénées. Visitors can pause, look around them, meditate and reflect, in this space that is free of charge and accessible to all. From the pathway, the visitor arrives at the entrance and discovers a silent monument, aligned with block F. Access to the memorial is indirect, via a ramp that is partially buried in the ground, thereby sanctifying the megalith, and becoming the stepping stone to a journey through time. This tunnel ends abruptly: the visitor finds himself facing a block that is 240 meters long, opaque and timeless, just a few meters from where he stands. After two strides in the daylight, the visitor enters a building where he will soon discover that

the only views or openings towards the exterior are towards the sky itself. The lobby is enveloped in a soft lighting and a calm and serene atmosphere reigns. Opposite the lobby is a long wall, devoid of any elements or decor, in which a passageway is situated. It is a long passageway, relatively narrow. The temporary and permanent exhibition spaces are arranged around a large pillared hall, artificially lit from the ground, with large-sized images projected onto vertical concrete walls.

The outside path forms a loop around the museum, marking the end of the visit. The memorial offers no view of the exterior, except for the sky. However, microcosms are present here and there inside the building. Three patios structure the organization of the learning labs, social area and offices, all the while providing a certain sense of comfort. This project is rooted in acceptance. Acceptance of the block, its lines, its military geometry transformed into something alienating, and of course, its history. The Rivesaltes Memorial, compacted between earth and sky, between past and memory, is situated exactly in the present and in life itself. Its formal violence demonstrates the impossibility of forgetting.

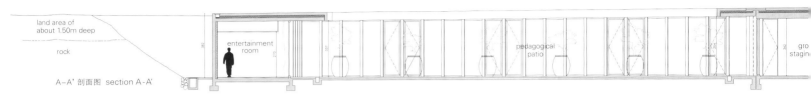

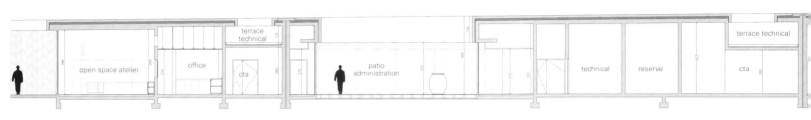

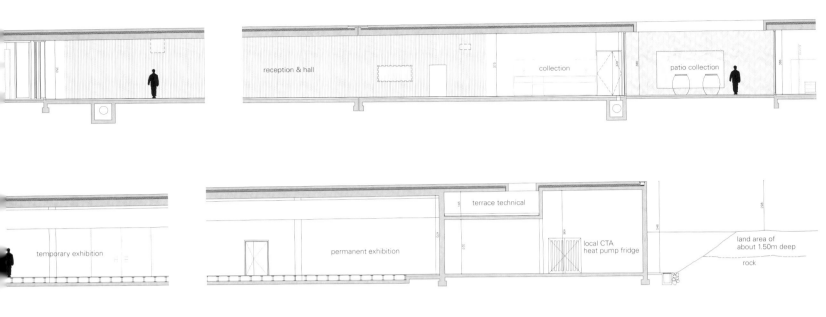

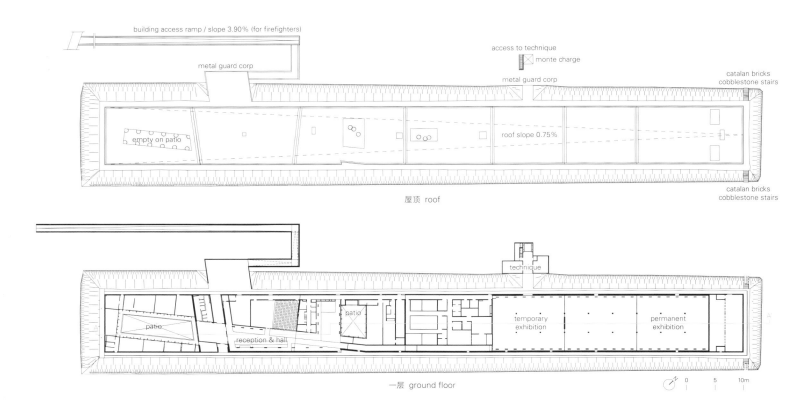

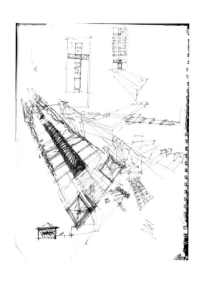 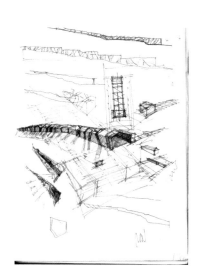 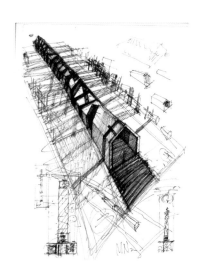

波兰村庄殉难者陵墓
Nizio Design International

波兰村庄殉难者陵墓位于Michniów，目前正在进行第五阶段的连续施工工程，陵墓的设计构思是建造一个整体的雕塑般的建筑形式，提供一个多媒体展览空间，通过建筑形状来表现以Michniów为代表的、在这儿曾发生的历史事件。该项目由著名的华沙建筑事务所Nizio Design International设计，计划于2017年建成开放。

波兰村庄殉难者陵墓项目受凯尔采地区农村博物馆委托建设，旨在纪念被德国占领期间的波兰农村社区的受害者，纪念他们所遭受的镇压。Michniów于1943年7月12日被德国采取"绥靖政策"，如今是所有那些第二次世界大战期间被绥靖的村庄的代表。现在陵墓项目所建的地方过去也被用来纪念那些曾发生的悲剧性事件：最初，修建了一个受害者集体墓穴（1945年），后来竖立了"Michniów圣母怜子雕像"，修建了国家纪念馆。

项目建设的第一阶段是完成场地的围墙、设备建筑和停车场，同时修筑多条通向集体墓穴的通道；第二阶段是建造地下结构部分；第三阶段是建造开放部分的基础；第四阶段，也是目前最困难的阶段，是建筑主体的建设。目前正在进行的是第五阶段，修建从集体墓穴到殉难者陵墓的道路。

这一巨大的整体结构设计充分应用了防水混凝土的连接和浇筑技术。这些钢筋混凝土壁柱从平板基础开始向上，在屋脊处拼接到一起。这些钢筋混凝土壁柱支撑起建筑的外表层，既是立面，又是保温层。在墙体和屋顶斜坡中，建筑师"隐藏"了通风管道、电信和电力基础设施等建筑元素。Mirosław Nizio工作室的设计设想是，钢筋混凝土构件的可见表面将复制木材的结构特征。为了达到这种效果，也为了修复钢筋混凝土结构，就在这些钢筋混凝土壁柱可见表面上薄薄地涂了一层特殊石膏——使用矩阵图案。这种矩阵图案的木材纹理效果被应用于屋顶的斜坡、内墙和侧立面上。

该建筑的特点是分段式结构。一道道裂缝将整个建筑组织横向切开，分成或封闭或开放的各个部分。这种形式就是雕塑般建筑的灵感和建筑师考虑建筑与其所表现的历史事件的一致性的合力。此外，由于场地有10%～15%的坡度，因此需要调整建筑的形状来适应场地。而接下来的封闭和开放部分——五个封闭部分和六个开放部分——引导参观者观看波兰村庄被绥靖的历史的展览，展示了其发展的各个阶段以及不断升级的镇压过程。随着这一历史事件的不断发展，与之相呼应，建筑也呈现出变形和"毁坏"，象征性地述说着在这里所发生的毁灭性灾难。各封闭部分之间、墙体的缝隙和屋顶都采用了玻璃，成为建筑特色。项目的总体表面积超过16 200m²，核心展区将占用1700m²，而临时展区将占用270m²。

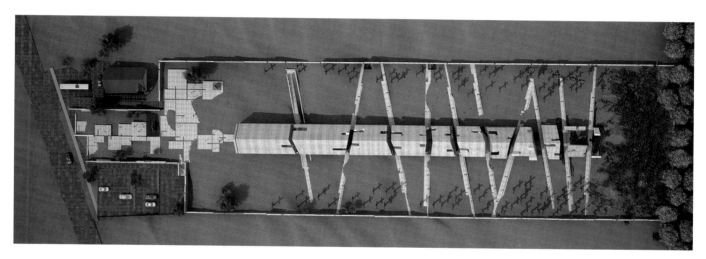

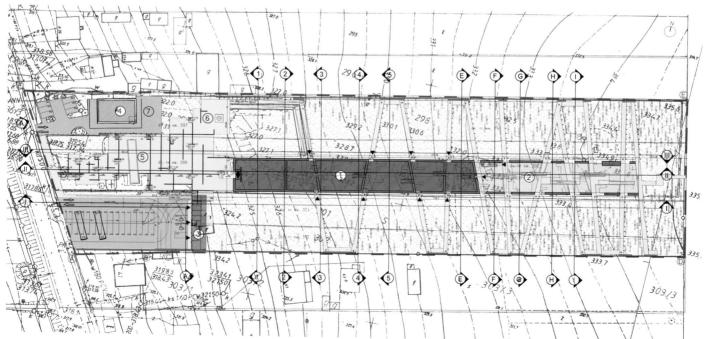

Mausoleum of the Martyrdom of Polish Villages

The Mausoleum of the Martyrdom of Polish Villages in Michniów is undergoing the successive fifth stage of construction works. The Mausoleum's design envisaged a monolithic sculptural architectural form to give room to a multi-media exhibition which through its shape is to convey the dramaturgy of the historical developments symbolised by Michniów itself. The Mausoleum, designed by the renowned Warsaw-based studio Nizio Design International, is scheduled to be opened in 2017.

The Mausoleum of the Martyrdom of Polish Villages is a project commissioned by the Museum of Kielce Region Countryside with a view to commemorating the victims of the repressions suffered by Polish rural communities during the German occupation. Michniów – pacified on 12 July, 1943 – today is a symbol of pacifications of villages that took place during World War II. The site on which the Mausoleum is being built used to be a symbolic place of remembrance dedicated to those tragic incidents: originally, a collective grave of the victims was erected (in 1945), followed by the "Pieta of Michniów" sculpture and National Remembrance House.

The first stage of works on the construction site of the Mausoleum delivered the site fence, services building, and the car park. Also, a number of paths leading to the grave were built at the time. Phase two involved building the underground structures, phase three produced the foundation of the open parts, and phase four, the most difficult so far, resulted in the

项目名称：Mausoleum of the Martyrdom of Polish Villages
地点：Michniów, Poland
建筑师：Mirosław Nizio _ Nizio Design International
建筑面积：16,200m²
核心展区面积：1,700 m²
临时展区面积：270m²
设计时间：2009 _ first prize in the competition for the architectural concept
竣工时间：2017 _ scheduled openinig of the Mausoleum
摄影师：©Lech Kwartowicz (courtesy of the architect)

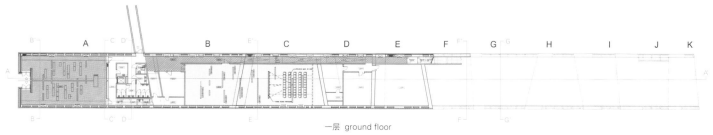

一层 ground floor

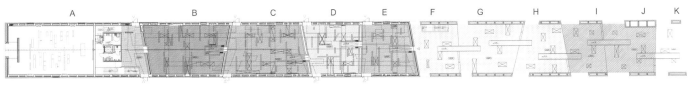

二层 first floor

construction of the essential components of the building. Currently underway is phase five whose objective is to build the approaches from the grave to the Mausoleum building. The object's monolithic structure was designed in full applying the watertight concrete jointing and injection technology. The ferroconcrete pilasters that rise from the slab foundation come together at the roof ridge. They support the external envelope that serves both as the elevation and the insulation layer. In the walls and roof slopes the designers "hid" such elements as ventilation ducts, telecommunication and electrical infrastructure. The design created by Mirosław Nizio's studio envisaged that the visible surfaces of ferroconcrete elements would reproduce the structural features of wood. This effect was achieved by applying on the surface – using matrix impressions – a special thin-layer plaster for ferroconcrete structure repairs. Such impressions were applied on the roof slopes, inner walls, and side elevations.

The building has a characteristic segmented structure. Its tissue is cut across by cracks that divide the architectural form into closed and open parts. This form is the resultant of the sculptural inspirations and thinking of the architecture's consistency with the historical narrative. Also, it was informed by the need to adjust the building to the shape of its site featuring 10%-15% slope of the ground. The subsequent closed and open segments – there are five of the former and six of the latter – lead visitors through the exhibition that shows the history of the pacification and presents its subsequent stages and the process of escalating repression. In parallel to the narrative the building undergoes deformation and "destruction", which symbolically conveys the annihilation that took place here. The gaps between the subsequent closed segments, the walls and the roof boast glass architectural features. The overall surface of the layout exceeds 16,200 sqm. The core exhibition of the projected museum will take 1,700 sqm, while temporary exhibitions will occupy 270 sqm.

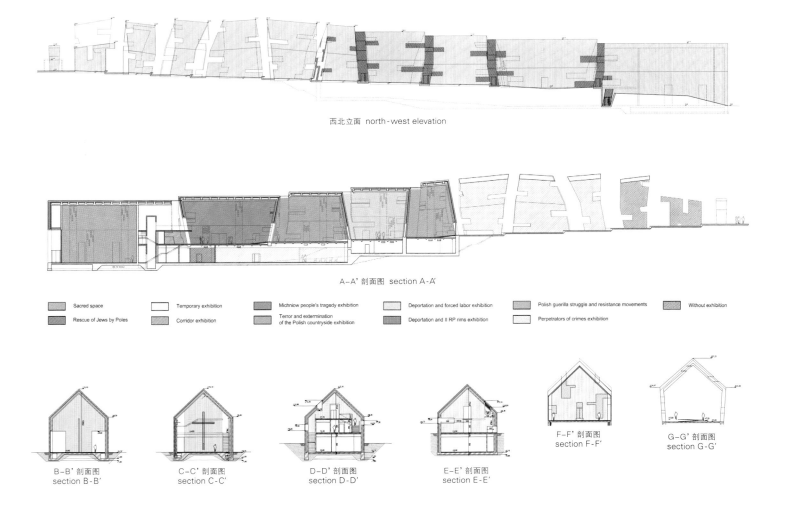

西北立面 north-west elevation

A-A' 剖面图 section A-A'

- Sacred space
- Rescue of Jews by Poles
- Temporary exhibition
- Corridor exhibition
- Michniow people's tragedy exhibition
- Terror and extermination of the Polish countryside exhibition
- Deportation and forced labor exhibition
- Deportation and II RP rims exhibition
- Polish guerilla struggle and resistance movements
- Perpetrators of crimes exhibition
- Without exhibition

B-B' 剖面图 section B-B'

C-C' 剖面图 section C-C'

D-D' 剖面图 section D-D'

E-E' 剖面图 section E-E'

F-F' 剖面图 section F-F'

G-G' 剖面图 section G-G'

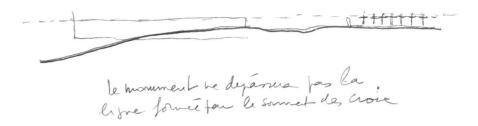

回忆之环

Agence d'Architecture Philippe Prost

2011年，法国北部加来海峡大区为了纪念第一次世界大战100周年，推出了一个国际纪念碑的建筑设计竞赛。选定的设计项目将建在洛雷特圣母山上，洛雷特圣母山是一个自然和遗产保护区，旁边是法国20世纪国家墓地。这个纪念碑需要在很短的时间周期内建成，并且预算紧张，需要精打细算。

建筑师设计的项目旨在战胜第一次世界大战给人们带来的恐惧，纪念参加第一次世界大战的退伍军人，从而满足了富有鲜明政治色彩的竞赛任务要求。目的是强调未来和平愿景对欧洲的重要性。

纪念碑为椭圆形环体，外围长328m，其黑色的纤维增强混凝土外观代表着战争的色彩和情绪，与俯瞰阿图瓦平原的洛雷特圣母山平衡而和谐。环体内部采用镜面品质的500张镀金金属片装饰，用来反射自然光线。

所有在北部加来海峡大区的战场上牺牲的579 606名士兵的名字都被刻在不锈钢钢板上，名字是按字母顺序排列的，不分国籍、等级或信仰，共同的人性将他们永远连为一体。

环形设计源于一种渴望，团结那些曾经的敌人。当人们相互牵起手来，就形成了圆环。

这个圆环是团结和永恒的代名词：团结，因为所有的名字构成了某种链条，将人与人连为一体；永恒，因为这些名字会一直循环往复，没有尽头。根据场地情况，圆环采用椭圆形，一侧朝向入口墓地，另一侧是阿图瓦平原。

纪念碑选择水平放置的原因似乎是不言而喻的，建筑师不仅要考虑原有的建筑——50m高灯塔的垂直度，而且认为水平状态是平衡和永恒的标志。三分之二的圆环部分扎根于大地，但随着山坡越来越陡，圆环就脱离了地面：这60m的悬臂提醒我们，和平将永远是脆弱的。为了实现上述设计，有必要使用后张预应力混凝土结构。该纪念碑在天空和大地之间创造了一个失重空间。

在这个地点，曾发生过可怕的战斗，现在大自然已收回她的权力；纪念碑将铭刻对那些在这儿倒下的士兵的记忆，此外，赞美歌颂失而复得的和平。

回忆之环是一个名副其实的艺术作品，一个技术上的挑战，一件不朽的作品，巧妙地与自然一较高下。超高性能的纤维增强混凝土这一新材料的使用使所设计的结构成为可能，任凭时间流逝，终将不朽。

The Ring of Remembrance

In 2011, the Region Nord-Pas-de-Calais launched an architectural competition for the construction of an international memorial celebrating one century of the First World War. The selected project was to be constructed on the Notre-Dame-de-Lorette hill, a natural and heritage-protected site alongside the French national necropolis of the twentieth century. This memorial site was to be built within a short period of time and had to accommodate a tight budget.

In order to respond to this strong political statement, we conceived a project that seeks to overcome the horror of the First World War and commemorates its veterans. The purpose is to emphasize the importance of a peaceful vision of the future to Europe.

The exterior, a 328 meters ribbon of dark fiber concrete representing the color and emotion of war, balances on the hill overlooking the plains of Artois. The interior uses the mirrorlike quality of 500 sheets of gilded metal to reflect natural light. The names of all of the 579,606 soldiers who fell on

A-A' 剖面图 section A-A'

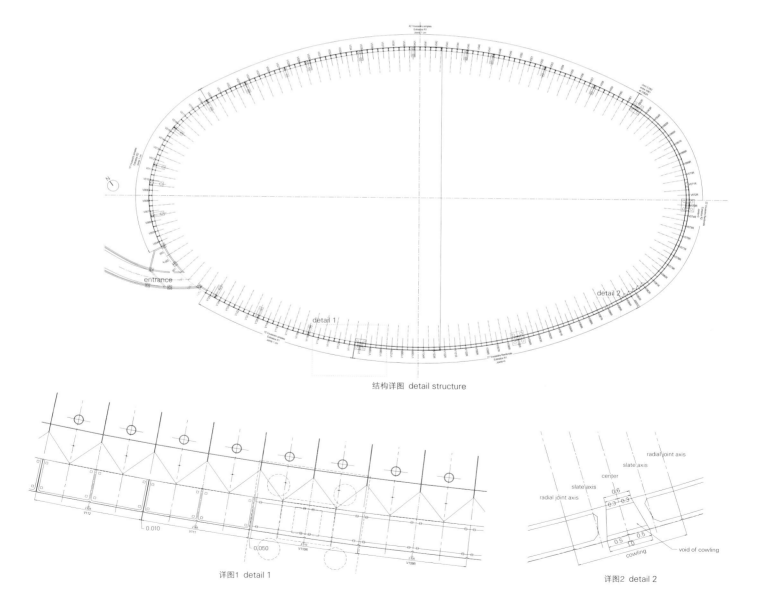

结构详图 detail structure

详图1 detail 1

详图2 detail 2

the battlefields of the Nord-Pas-de-Calais are inscribed on the stainless steel sheets, are arranged in alphabetical order with no distinction of nationality, rank or creed, to be forever united in their common humanity.

The form of the ring was influenced by the desire to unite those who were once enemies, keeping in mind the circle that is formed when people hold hands.

The ring is synonymous with both unity and eternity: unity because the names now constitute a sort of human chain; eternity because the names continue without end. Due to the way the structure sits on its site, the ring takes the form of an ellipse, turned on one side towards the entrance of the necropolis and on the other towards the plain of Artois.

The choice of horizontality for the memorial appeared self-evident. Not only did we want to take into consideration the verticality of the previously existing 50-meter-high lantern tower, but also the idea that horizontality is a sign of balance and timelessness. With two-thirds of the structure's diameter

rooted in the ground, the ring detaches itself from the earth where the slope of the hill becomes steeper: this 60-meter cantilever reminds us that peace will always remain fragile. To achieve this, it was necessary to use prestressing by post-tensioning. The memorial creates a weightless space between the sky and the earth.

On the same site where horrific battles were played out nature has now reclaimed her rights; the memorial will inscribe the memory of the fallen through space and will moreover celebrate rediscovered peace.

The ring is a work of art in every sense of the term: a technical challenge, a monumental work and an artifice competing with nature. The use of a new material, an ultra high performance fiber-reinforced concrete, has made its construction possible and will allow it to defy the passing of time.

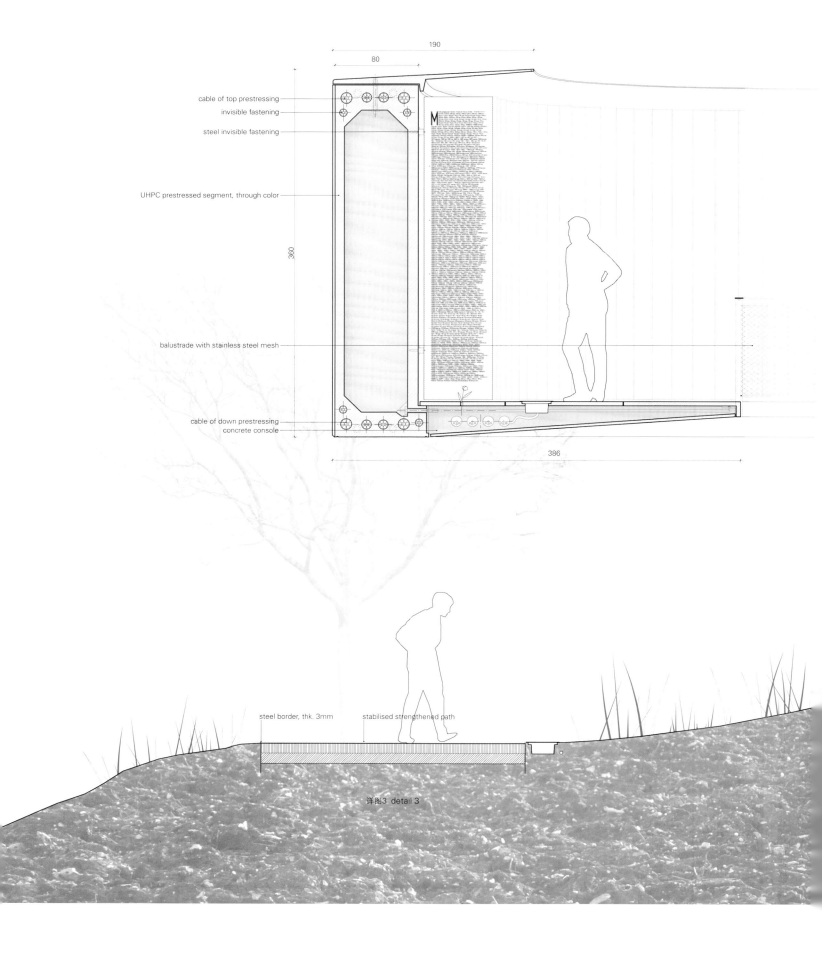

详图3 detail 3

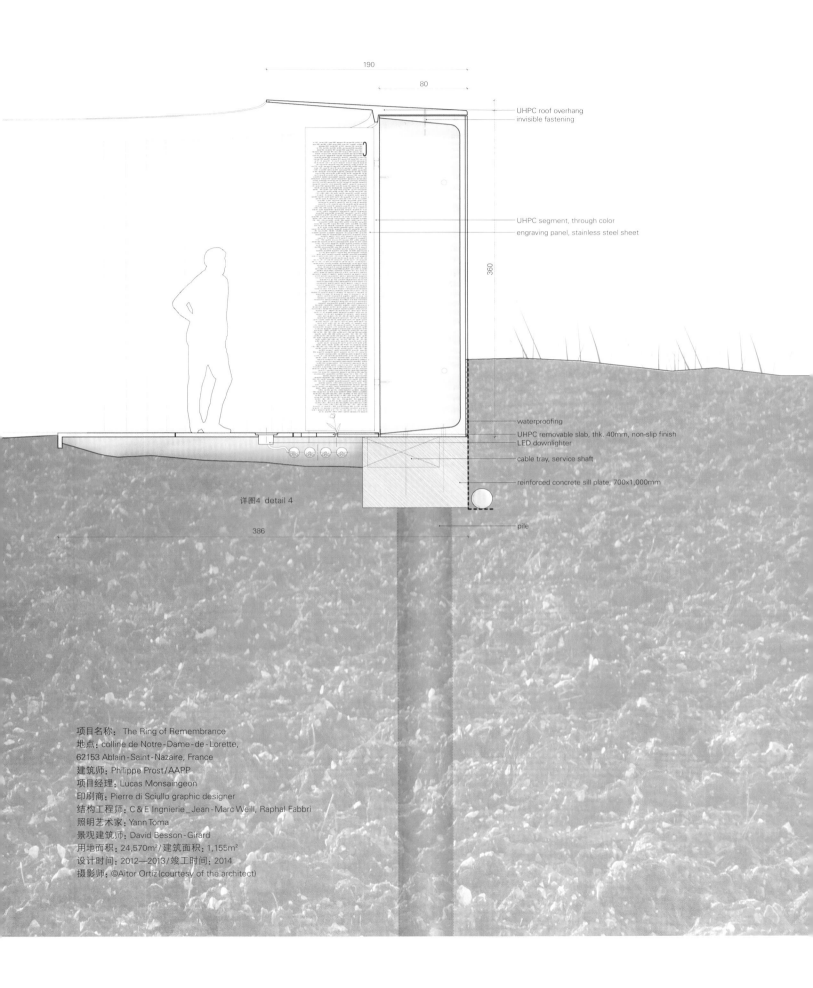

详图4 detail 4

项目名称：The Ring of Remembrance
地点：colline de Notre-Dame-de-Lorette,
62153 Ablain-Saint-Nazaire, France
建筑师：Philippe Prost/AAPP
项目经理：Lucas Monsaingeon
印刷商：Pierre di Sciullo graphic designer
结构工程师：C & E Ingnierie _ Jean-Marc Weill, Raphal Fabbri
照明艺术家：Yann Toma
景观建筑师：David Besson-Girard
用地面积：24,570m² / 建筑面积：1,155m²
设计时间：2012—2013 / 竣工时间：2014
摄影师：©Aitor Ortiz (courtesy of the architect)

建筑情感：从宗教到世俗 Architecture of Sentiment – from the sacred to the human

博洛尼亚大屠杀纪念碑
SET Architects

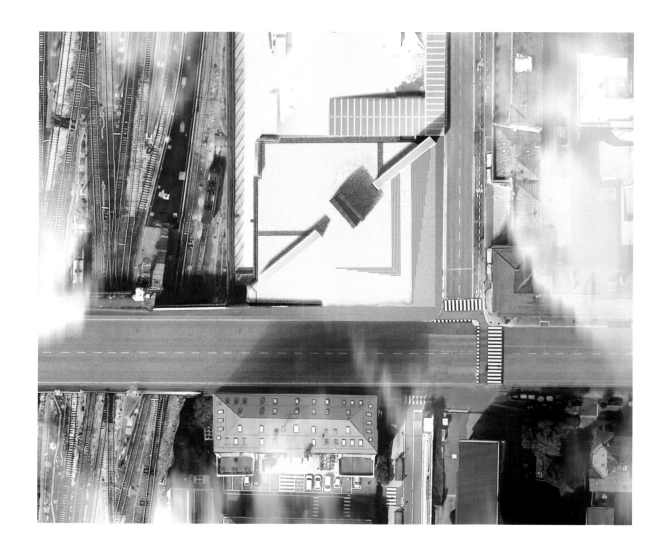

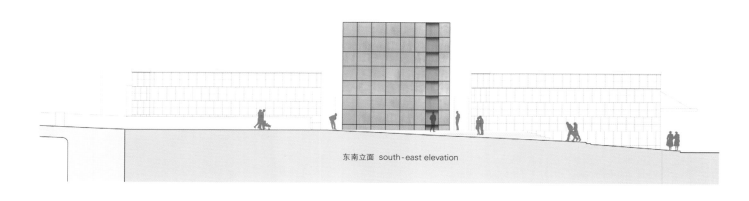

东南立面 south-east elevation

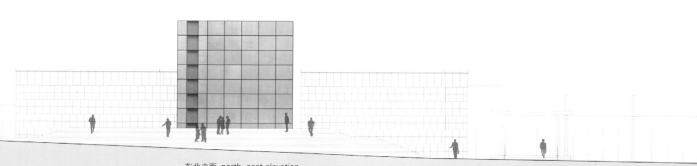

东北立面 north-east elevation

博洛尼亚大屠杀纪念碑用了不到两个月的时间就建好了，被赋予了巨大的情感力量，是一个非常显眼的地标，位于Via dé Carracci街和Ponte Matteotti街交会处的城市广场，被新建的博洛尼亚高铁站所环绕。这一地区将成为城市新的连接纽带。这样的话，纪念碑会吸引行人，使他们驻足，思考大屠杀这一悲剧。

纪念碑由两个对称的10m×10m考顿钢平行六面体组成，两个六面体并列在一起，与广场上的墙垂直。两个并置的六面体之间形成一条通道，最开始有1.60m宽，但越来越窄，最后只有80cm宽，立即给人带来一种压迫感。

在纪念碑的内部，是横向和纵向的金属板90°相交形成的网格，构成一系列1.80m×1.25m的矩形空盒体——这些格子象征着集中营宿舍的牢房。纪念碑的外立面俯瞰着这座城市，就像一张白纸——仿佛它就是一段有待撰写的历史。另外，小格子外沿的钢板都稍稍突出，象征着当代意识情感。

考顿钢的选择是经过深思熟虑的，考顿钢放置露天就会自然生锈。随着岁月的流逝，它的腐蚀会显示时间的痕迹，表明所有的事情背后都有一段丰富的历史。两个六面体之间的通道由碎石铺设，是铺设路基路面所使用的典型的玄武岩碎石。这条通道代表"月台"，代表奥斯威辛和比克瑙集中营之间的铁路：德国人利用铁路运来了无数的犹太人、波兰人、吉普赛人，等等。

当人们走过这条碎石通道，脚步发出空荡荡的回响，再加上通道的狭窄，使人们渐渐产生一种强烈的痛苦感。纪念碑以这样的方式来展现生命，以这样的方式来唤起深刻的回忆。此外，灯光起着至关重要的作用，为纪念碑的设计锦上添花。白天的时候，整个广场沐浴在阳光中，而通道里面的光线却非常暗淡，适合人们沉思，使游客能够冷静思考。然而晚上的时候，精心设计布置的人造光源照亮了主建筑体量，使纪念碑看起来更加威严。

总的来说，其历史抱负使该纪念碑独树一帜，为了强调情感的重要性而摒弃那些修饰和说教的传统做法。用这种方式，SET建筑师事务所成功设计了一个利用现在情感来讲述过去的纪念碑。

Bologna Shoah Memorial

Built in less than two months, the Bologna Shoah Memorial is a recognizable landmark of great emotional power. It is located at the intersection of Via dé Carracci and Ponte Matteotti, a city square encompassed by the newly installed high-speed train station of Bologna. This area is primed to become the new connective pole of the city. As such, the monument attracts passers-by, inviting them to reflect on the tragedy of the Holocaust.

The Memorial is made up of two symmetrical corten steel parallelepiped blocks of 10 x 10 meters each; the blocks sit adjacent to one another, perpendicular to the existing walls of the square. Their position converges to create a path, which begins with a width of 1.60 meters, and drastically narrowing to just 80 centimeters. The path generates an immediate feeling of oppression.

At the interior of the Memorial, the volumes present a grid of horizontal and vertical metal sheets which intersect at 90°, giving shape to a series of rectangular empty boxes of 1.80 x

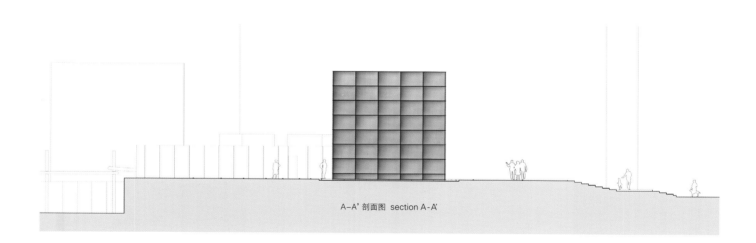

A-A' 剖面图 section A-A'

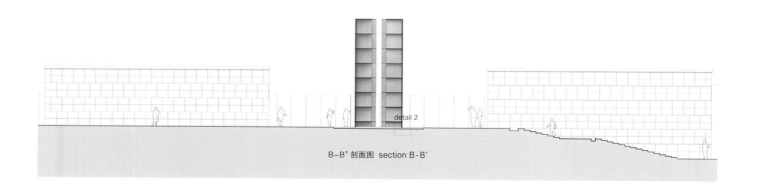

B-B' 剖面图 section B-B'

项目名称：Bologna Shoah Memorial / 地点：Via Giacomo Matteotti / Via dè Carracci, Bologna, Italy / 建筑师：SET Architects / 项目团队：Lorenzo Catena, Chiara Cucina, Onorato di Manno, Andrea Tanci / 结构工程师：Proges Engineering - Ing. Andrea Imbrenda / 总承包商：Sì Produzioni / 客户：Comunità Ebraica di Bologna
金属加工：Officina Paolo Cocchi / 地面铺设：Edil Nuova S.A.S. / 类型：Erco / Type: Memorial / 用地面积：2,000m² / 建筑面积：40m² / 设计时间：2015 / 竣工时间：2016
摄影师：©Simone Bossi (courtesy of the architect) - p.138~141, p.144~147, p.150~153, ©Visual Lab Bologna (courtesy of the architect) - p.148 bottom-left, p.149 bottom

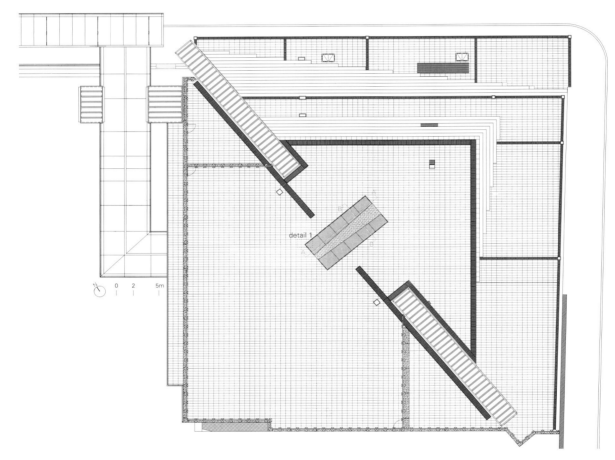

detail 1

0 2 5m

详图1 detail 1

- existing paving 3cm
- steel slab CORTEN 8mm
- welding
- steel drainage grate
- steel slab CORTEN 8mm
- new basalt stone paving

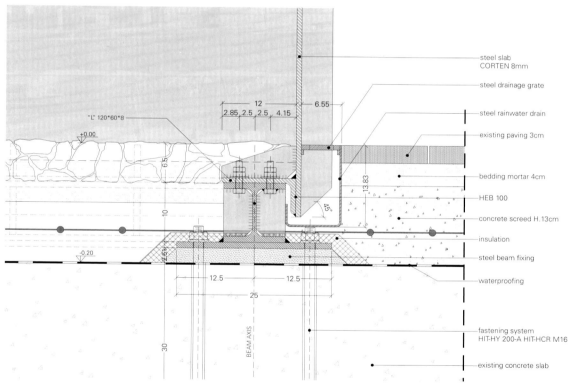

- steel slab CORTEN 8mm
- steel drainage grate
- steel rainwater drain
- existing paving 3cm
- bedding mortar 4cm
- HEB 100
- concrete screed H.13cm
- insulation
- steel beam fixing
- waterproofing
- fastening system HIT-HY 200-A HIT-HCR M16
- existing concrete slab

详图2 detail 2

1.25 meters – these boxes represent the cells of the dormitories in the concentration camps. The exterior facade of the Memorial overlooks the city, resembling a blank page – as if it is of a history yet to be written. And, along the perimeter of the cells, slight steel protrusions symbolize feelings of contemporary awareness.

The choice of corten steel is deliberate: it is a material that will naturally rust when exposed to open air. As the years pass its corrosion will display the vestiges of time, demonstrating that all things have a rich history behind them. The paving of the path between the two blocks is realized in ballast, basalt stone chippings typical of the roadbeds. This represents the "Judenrampe", the railroad tracks between the Auschwitz-Birkenau camps where the Germans sent countless transports full of Jews, Poles, Roma and others.

The empty echoes of footsteps across the stones coupled with the restriction of the passage instills a keen sense of anguish: in this way the Memorial takes on life and evokes the drama of the memory.

Further, light plays an essential role in the culmination of the monument. During daytime when the square is lit by sun's rays, the passage becomes immersed in a dim, contemplative light, allowing visitors to calmly reflect. Then at night, strategically placed artificial light illuminates the primary volumes, magnifing the majesty of the Memorial.

In total, the Memorial, distinguished by its historical ambition, abandons rhetorical and didactic conventions in order to emphasize the importance of emotions: in this way SET Architects succeeded in designing a monument that utilizes present sensibility to narrate the past.

麻省理工科利尔纪念碑

Höweler + Yoon Architecture

科利尔纪念碑坐落于麻省理工学院的校园内，是为纪念在2013年4月18日中枪身亡的肖恩·科利尔校警而建造的。科利尔纪念碑这一永恒建筑成为这个悲剧发生地的标记，而"威武的科利尔"这一词语被转化成一个充满回忆的空间，其造型体现了团结所赋予人们的力量。

纪念碑由32块花岗岩石块组成，形成一个五向石板穹顶。石块之间相互支撑构成一个中空拱形结构，令人沉思。纪念碑的设计灵感源于张开的手掌，薄薄的石板穹顶由五个径向壁支撑，向外伸向校园。径向壁中央的卵形空间设有一条通道，还有纪念碑的标志。这个孔径重新定义了这个地方。星形结构的交叉点和中央空间形成一个光滑曲面，曲面的下面是纪念碑的标志，上面写着"2013年4月18日肖恩·科利尔因公殉职"。

纪念碑较长的两面墙像围挡一样面向瓦萨街，就像是在保护这个地方，同时也自然而然地成为纪念碑的入口。形成夹角最小的两面墙正对着发生枪击的地点，只有几英尺远。朝向南面的墙上刻着墓志铭，是肖恩哥哥的悼词："愿永垂不朽。宽广的胸怀，满面的笑容，周到的服务，博爱众生。"一簇簇皂荚树在这一坚固的石头建筑上形成生机勃勃的华盖，记录着时间的流逝。而嵌入步道地砖里的点光源则永远铭记下了2013年4月18日那天夜空的星座。

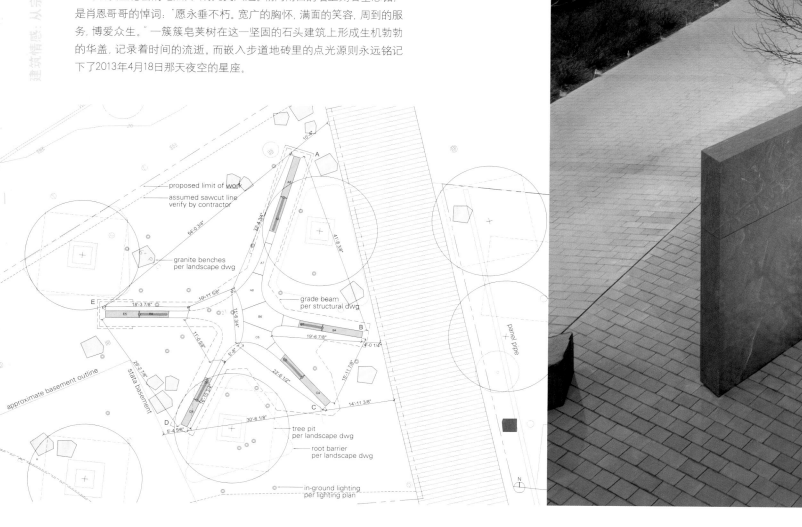

MIT Collier Memorial

纪念碑的设计将古老的横跨砌筑拱顶建筑技巧和新的数字制造与结构计算技术结合起来，创造出前所未有的建筑形式。石板拱是整个结构构造中最重要的元素，需要合理安排构建空间的建筑材料，将力转换为形式。整个设计依赖于32块花岗岩石块的精确安放，通过石块之间的纯压缩来传递荷载。空中轻薄而巨大的石板穹顶营造了悬浮和失重的效果，而形成整个压缩环的每一个花岗岩石块的锥形几何结构却是这个石头拱形建筑最重要的几何结构。

石块的加工过程包括对开采的石块的切割，先用单轴机器人块锯进行切割，再用多轴机器人库卡500进行切割。拱顶的几何结构要求石板间必须严丝合缝，机器人对石块的铣削过程要求最后加工出来的石块与数学模型之间的误差不超过0.5mm。从技术方法上来说，肖恩·科利尔纪念碑的设计过程包括在实物模拟数字模型的建造和用数字工具进行模拟实验之间来回反复这样的过程。然后，泥瓦匠搭建顺序极为错综复杂的脚手架，现场安装这些巨大的石块。

这一设计不仅展现了新的数字建造方法，也展现了传统的石块砌筑技艺，既赞美了当代技术，也颂扬了时间恒久远的手工艺。肖恩·科利尔纪念碑的拱形设计体现了建筑材料排列布局的结构原理，象征着慷慨的服务。这种视觉化的说教力量与麻省理工学院开放透明的精神是一致的。而要实现这样一个稳定的形状，所有五个径向壁缺一不可，这象征着整个社区共同纪念所遭遇的损失（译者注：肖恩·科利尔校警的牺牲）。永恒的科利尔纪念碑让人们永远记住肖恩·科利尔警官，纪念他的生命和无私的服务，同时也表现了人们共同的价值观：面对威胁时所表现出的率真，从多样化中寻求统一以及社区所赋予人们的力量。

Situated on MIT's campus in honor of Officer Sean Collier who was shot and killed on April 18th 2013, the Collier Memorial marks the site of tragedy with a timeless structure – translating the phrase "Collier Strong" into a space of remembrance through a form that embodies the concept of strength through unity.

The memorial is composed of thirty-two solid blocks of granite that form a five-way stone vault. Each block supports others to create a central, covered space for reflection. Inspired by the gesture of an open hand, the memorial's shallow stone vault is buttressed by five radial walls, which extend outward toward the campus. The ovoid space at the center of the radial walls creates a passage, a marker, and an aperture that reframes the site. The intersection of the star-shaped form and the central void creates a smooth, curved surface on the underside, which acts as a bevel marker and reads "In the line of duty, Sean Collier, April 18, 2013".

The longest walls of the memorial shelter the site from Vassar Street and simultaneously create an entry into the memorial.

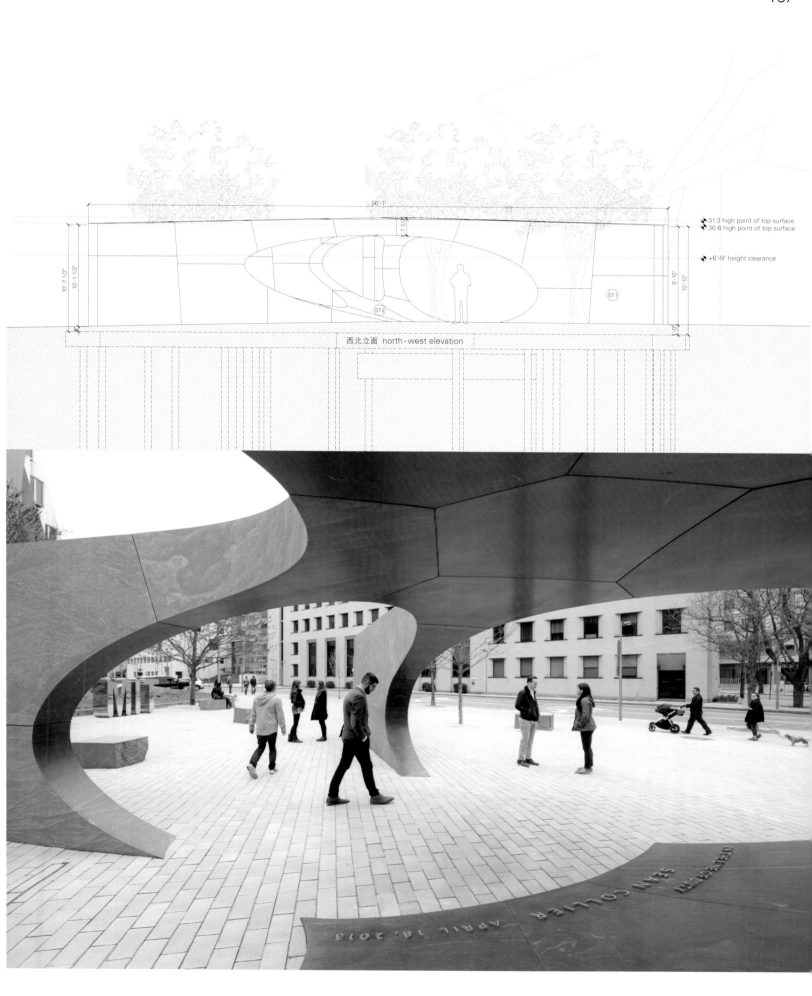

The two most acute walls are aligned with the location of the shooting just a few feet away. Carved into the south-facing wall is an epitaph from Sean's brother's eulogy, "Live long like he would. Big hearts, big smiles, big service, all love." Clusters of honey locust trees create a living canopy above the solid stone structure to mark the passage of time. In contrast, point lights set into the pavers permanently inscribe the constellation of stars in the sky the night of April 18th, 2013.

The design of the memorial combines age-old structural techniques for spanning masonry vaults with new digital fabrication and structural computation technologies to create an unprecedented form. The stone arch is among the most elemental of structural organizations, ordering materials in space and translating force into form. The design relies on the exact fit of the 32 stone blocks to transfer loads in pure compression from stone to stone. The shallowness of the massive stone vault overhead creates an effect of suspension and weightlessness, while the tapered geometry of the individual stone blocks that form the compression ring reveals the keystone geometry of the masonry arch.

The stone fabrication process involves the cutting of quarried blocks of stone, first with a single-axis robotic block saw, then with a multiple axis KUKA 500 robot. The vault geometry necessitates a perfect fit between blocks, and the robotic milling process produces final stone pieces that are within a 0.5 millimeter tolerance of the digital model. Methodologically, the design process for the Sean Collier Memorial involved a back and forth process between the construction of physical, analog, and digital models and simulations with digital tools. The massive stone blocks are then set on-site by masons through an intricate scaffolding sequence.

The design showcases both new digital fabrication methods as well as traditional stone setting masonry techniques, celebrating both contemporary technology and timeless craft. The vaulted design of the Sean Collier Memorial embodies structural principles in its material configuration and symbolizes generosity as service. This didactic visualization of forces is consistent with MIT's ethos of openness and transparency, while the idea that all five walls are needed to achieve a stable form is symbolic of a community coalescing to commemorate a loss. The permanent Collier Memorial offers the opportunity to remember Officer Sean Collier and honor his life and service, and represent shared values: openness in the face of threat, unity through diversity, and strength through community.

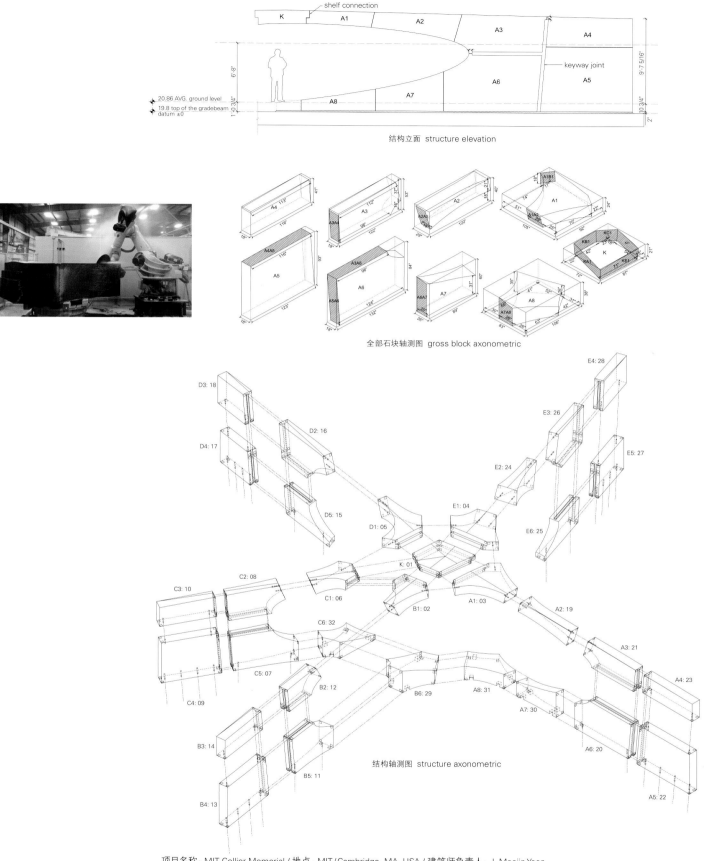

结构立面 structure elevation

全部石块轴测图 gross block axonometric

结构轴测图 structure axonometric

项目名称：MIT Collier Memorial / 地点：MIT/Cambridge, MA, USA / 建筑师负责人：J. Meejin Yoon
记录建筑师：Höweler + Yoon Architecture (J. Meejin Yoon, Eric Höweler, Yoonhee Cho, Paul Cattaneo, Sung Woo Jang, Anna Kaertner) / 结构工程师：Knippers Helbig Advanced Engineering / 特殊砌体顾问：Ochsendorf DeJong and Block Consulting Engineers / 客户：MIT / 功能：permanent memorial, campus gateway / 记录工程师：RSE Associates
景观建筑师：Richard Burke Associates / 土木工程师：Nitsche Engineering / 岩土工程师：McPhail Associates
照明设计师：Horton Lees Brogden Lighting Design / 电气工程师：AHA Consulting Engineers / 石材制造商：Quarra Stone Company / 花岗岩安装：Phoenix Bay State Construction Company / 花岗岩开采：Virginia Mist Group / 用地面积：603.87m²
设计时间：2013 / 竣工时间：2015 / 摄影师：©Iwan Baan (courtesy of the architect)

建筑情感：从宗教到世俗 Architecture of Sentiment – from the sacred to the human

日本石卷市石头纪念碑
Koishikawa Architects

2011年3月11日，发生了东日本大地震。这次地震对日本东部的大部分区域造成了大范围破坏。石卷市石头纪念碑坐落在受地震影响地区的地理中心位置，为人们提供了一个为受灾地区所有遇难者祈祷的场所。

石卷市石头纪念碑矗立在宫城县石卷市的一个小山坡上，人们可以在此为近18 000名在这次大地震中失踪和死亡的人祈祷。层层石瓦的数量代表日本的整个东部地区，也代表遇难者的人数，访客可以在此追忆这些遇难者。纪念碑由当地出产的石材和镜面不锈钢建造，其形状明确暗示了受灾最严重的地区在哪个方向，因此，访客自然而然地面向受灾最严重的地区祈祷。设计师由层层石瓦砌筑这个空间的目的也是让人们记住这个重大灾难，并代代相传。此外，建造纪念碑使用的金属板是镜面不锈钢板，可以映射周围的樱桃树。樱花开花的季节就会让人想到东日本大地震，就会让人想到这次大地震给人们的经验教训，从而为重建和再生提供了一个机会。

该项目使用当地的材料和施工方法建造。尽管最初石卷市的Ogatsu Tome石是被用来制作砚台和铺设屋顶的材料，但是当地工业已被地震摧毁。作为区域重建工作的开始，这种石头就被用来建造纪念碑。然而，由于当地经济尚未恢复，开采新材料很难，所以建造该项目所需要的石材是重新利用了从地震受损建筑上所拆下的旧的石材。从屋顶拆下的材料被挑选分类。铺设屋顶的石板无论状态好坏都被有益地利用起来，有效地支撑着纪念碑墙壁的小荷载。通过利用本地生产的材料，并将建筑材料与建筑形状很好地结合起来，这个空间的意义延伸了地方性，体现了所有受地震影响的地区。

此外，这种材料不需要任何重型设备来施工，项目本身就是一个小单元，一切可以用手运输和操作。更具体地说，Inai石、Ogatsu Tome石和基础被重新利用，表面采用了日本常见的Tataki饰面。

Stone Memorial in Ishinomaki

March 11, 2011, The Great East Japan Earthquake has occurred. The earthquake, has resulted in extensive damage to a wide area of east of Japan. "Stone memorial in Ishinomaki" is located in the geographic center of the affected areas to create a place that can offer prayers for all the victims of a wide devastated region.

"Stone memorial in Ishinomaki (Ishi-no-kinendo)", stands on a hillside in Ishinomaki, Miyagi, built as a space where people can pray for the nearly 18,000 people missing and dead. The number of laminated stones represent that of the victims which visitors can recall, as well as the entire east area of Japan. The shape of the prayer space, which was made of locally-produced stone and mirrored stainless steel, implies the heavily affected areas so explictly that visitors naturally stand and pray toward them. Laminated stone space is also aimed to remember the fact of this major disaster and inherit it to the future generations. In addition, the memorial plate

场地分析
Analysis of the site

Wide area

Mid range

Near range

①Tombstone

②Steep hill

③Big cherry tree

行走路线
Walking path

- Slope

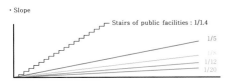

— 1/20 : Recommended standards for barrier-free law
— 1/12 : Limit criteria of barrier-free law
— 1/8 : Standards in the Building Standards Law
— 1/5 : Limit gradient electric wheelchair can pass

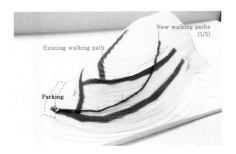

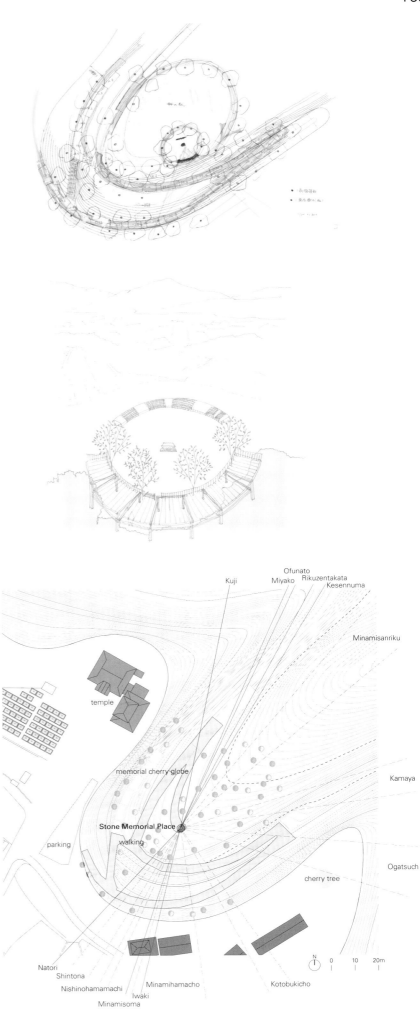

项目名称：Stone Memorial in Ishinomaki
地点：Miyagi, Japan
建筑师：Hiroya Kobiki + Noritaka Ishikawa _ Koishikawa architects
监理：Yoshiyuki Kawazoe _ Associate professor,
Kawazoe Lab., UTokyo
用途：memorial / 建筑面积：6.47m²
有效楼层面积：6.47m² / 建筑规模：1story
结构：stone
设计时间：2012.12~2014.8
施工时间：2014.9~2014.12
摄影师：©Koji Fujii

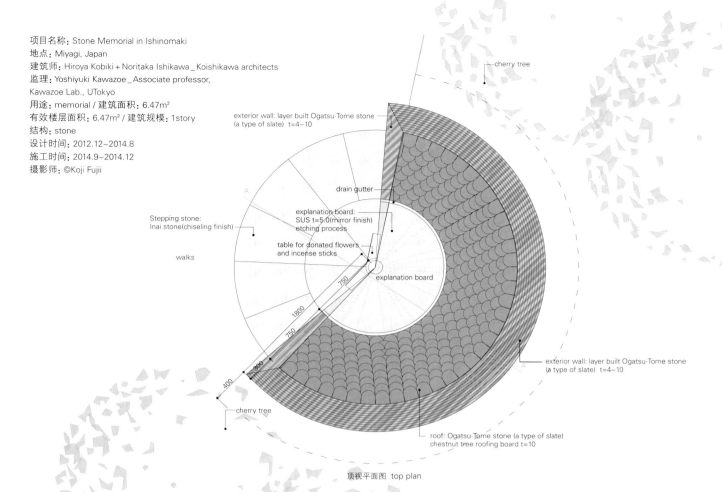

顶视平面图 top plan

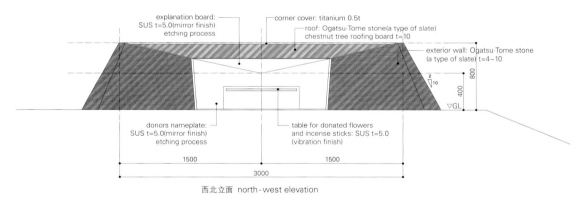

西北立面 north-west elevation

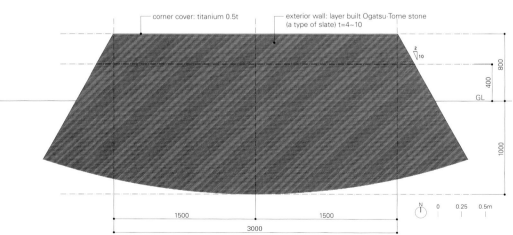

东南立面 south-east elevation

made of mirrored stainless steel, reflects the periphery of the cherry tree. The season when it blooms will bring to mind the Great East Japan Earthquake as well as the lessons learnt from it, offering a chance for reconstruction and regeneration. The project has been built using local materials and construction methods. Although originally Ogatsu Tome stone in Ishinomaki was used as an inkstone and roofing materials, local industry was destroyed by the earthquake. As an intial of regional reconstrcution, this stone was taken advantage as a memorial. However, since mining new materials was difficult due to not being yet resumed, the project was realized by reusing the waste generated by the demolition of existing building. Materials were sorted from dismantling roofs. Both roofing slates in good and bad conditions were valuably used and the small load of the wall was effectively supported. By taking advantage of locally produced material and incorporating the shape of the building with it, the meaning of the space extends placeness, and embraces all the areas affected by the earthquake.

In addition, the material did not require any heavy equipment for constructions and the project was planned as a small unit that can be transported and applied by hand. More specifically, Inai stone, Ogatsu Tome stone and the foundation were re-used with a finish not necessary Tataki, usually traditional in Japan. Koishikawa Architects

Integer pieces | Broken pieces

Slate roofing | Layer built

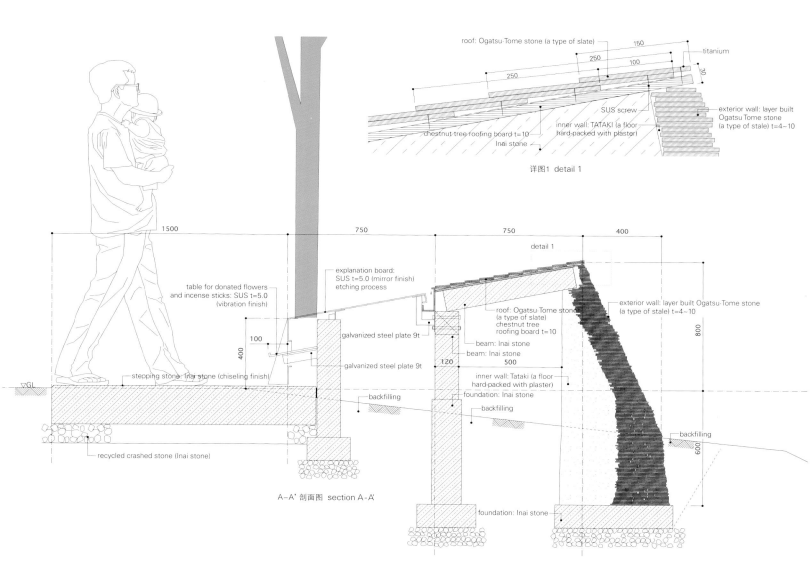

详图1 detail 1

A-A' 剖面图 section A-A'

建筑情感：从宗教到世俗 Architecture of Sentiment – from the sacred to the human

乌托亚纪念碑
3RW Arkitekter

于特岛是大自然中一处独特的所在,也曾是挪威历史上一处发生过最惨绝人寰的犯罪的地点。2011年7月22日,69人在此被残忍杀害,其中大多是孩子和年轻人。在这两个背景下,自然就代表着希望。在大自然环境中,我们人类开荒劈地,建立社会群落;建造房屋,为我们遮风挡雨。这一文化景观告诉人们人和社会如何改变了大自然,人们如何一代代适应大自然。

当森林中一棵大树死去时,一个有机过程也就开始了,就像我们所知道的留出了一块林中空地——在浓密的林中留出了空间。建筑师想要把于特岛上的纪念碑就建造成这样的一块林中空地,形成一个明确的地方,突出这个地方原有的所有美景,与周围环境融为一体。

建筑师的目的是把参天松树之间的空地修建成一个大的圆形,地形要比原有地势稍低,但还是向海面倾斜并保持圆形露天剧场/碗的形状。从于特岛中心建筑区域有一条通往纪念碑的小路。在小路与圆形空地交接处,建筑师将圆形空地边上的地势稍微降低一些,这样圆形空地的边缘可供访客就坐休息;在其另一端,地势稍稍抬起,这样就形成了另一个就座边缘,访客可以坐在上面,面朝大海和Sørbråten(岸上的公共纪念馆)。沿着圆形空地的边缘铺上了表面粗糙的石板,这样人们可以四处走动,即使坐着轮椅也可以。圆形空地上边和下边边缘也铺设了石板,人们坐在上面更加舒适一些。石板也可以防止新的植被和野草过于茂盛生长而破坏这处林中空地,使其不需要大量维护,进而保护这块空地。

在圆形空地中,建筑师修建了一个小花园,里面精心挑选种植了在于特岛上可以找到的能够吸引众多种类蝴蝶的植物种类。利用高大的松树悬挂了一个沉重的金属圆环,上面刻着所有遇害者的名字。阳光穿过圆环,可以照到每一个名字。访客围绕着圆环移动,可以读到每个人的名字。

所有的建筑工程都有志愿者的高度参与。精确度方面没有过多要求,其设计理念也与惨案发生地点当地的地势条件相适应。

Memorial at Utøya

Utøya is both a unique piece of nature, and the scene of one of the most horrific crimes in Norway's history. Sixty nine people, most of them children and youth, were brutally murdered here on the 22nd July 2011. Within this duality, nature represents hope. In this nature, we as humans have cleared space to establish social communities and have made shelter to protect us from the weather. This cultural landscape tells a story about how people and society have changed nature, and adapted to it through generations.

When a big tree in the forest dies, an organic process that creates what we know as a clearing begins – open room in the otherwise dense forest. We want to develop the memorial on Utøya as such a clearing, forming a clear spot and highlight all the beautiful landscape qualities that are already in place and in contact with the surrounding environment.

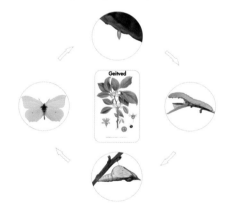

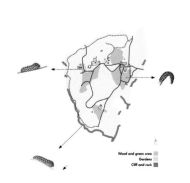
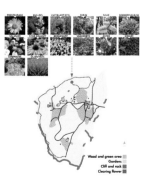
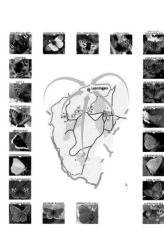

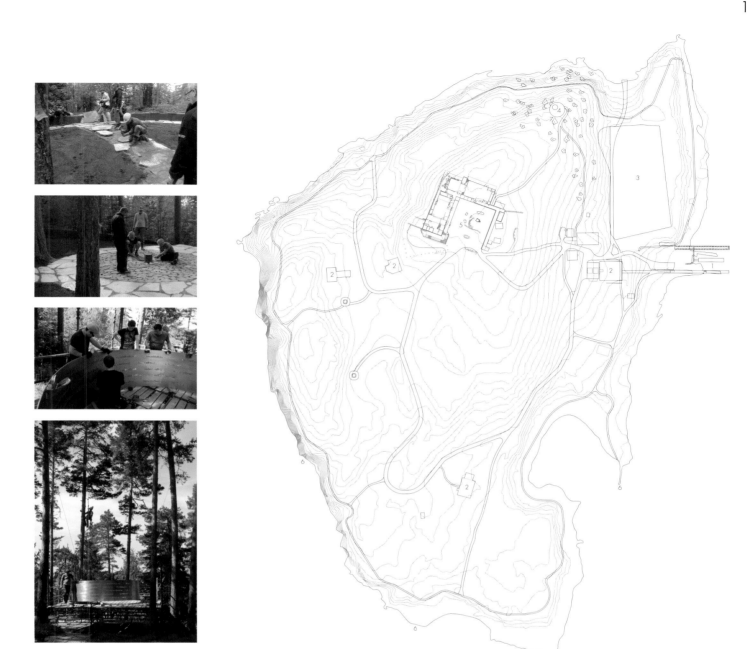

1. 到达与离开的出入口 2. 设备建筑 3. 体育设施区域 4. 乌托亚纪念碑 "The Cleraing"
5. 活动中心 6. 蝴蝶出生与蜕变区 7. 花朵吸引蝴蝶移动
1. Arrival and departure pier 2. Facility building 3. Sport facility area 4. Memorial at Utøya "The Cleraing"
5. Activity centre   6. Birth and transformation area for butterfly 7. Flower attracting butterfly migration

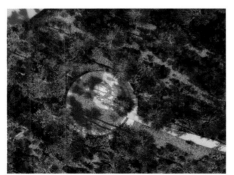
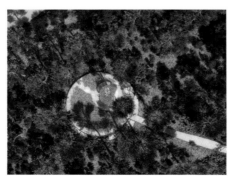

Our goal is to shape the open space between the big pine trees as a major unifying circle. The circle is set in the landscape with a slightly lower slope than the current terrain, but still allowing the sloping down towards the water and preserves an amphitheater/bowl shape. When the path from the central building areas at Utøya comes in to the memorial, we lower the terrain gently along the circle, thus creating a seating edge where you approach the circle. On the opposite side of the circle, the terrain is slightly lifted and you create in this way another edge where you are facing out of the circle, towards the water and Sørbråten (public memorial on the shoreside). Along the edge of the circle, we pave the ground with rough slate-stones. This will allow people to move around, even in wheelchairs. The top and bottom edges are also covered with slate so they are more comfortable to sit on. Slate slabs will prevent that new vegetation and weeds grow too aggressive, therefore deteriorating the clearing. It will thus conserve the clearing without requiring extensive maintenance.

Inside the circle we build a garden with specially selected plants that attract many species of butterflies found on Utøya. From the tall pine trees, we hang a heavy metal ring where the names of all victims are carved out. The names will read by the light that shines through the plate. Moving around the ring, one can read all the names.

All building work is planned so that it can be performed with a high degree of volunteer effort. There are few requirements for precision, and the concept can be adapted to local terrain conditions on site.

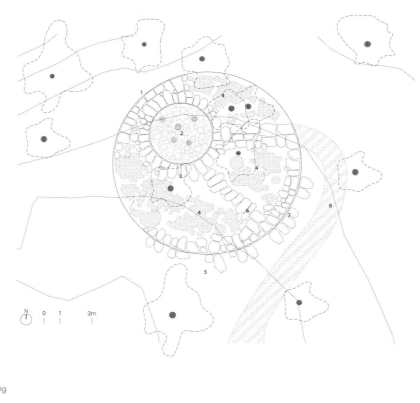

1. 面向水池与石板路的座位区
2. 树干构成的木地板
3. 直径4m的金属环
4. 蝴蝶花种植区
5. 直径12m考顿钢构成的景观区
6. 石板路第二入口
7. 面向金属环与石板路的座位区
8. 砾石路
9. 固定金属环的树
10. 面向金属环、石板路和考顿钢带的座位区

1. seating area toward the water with slate paving
2. wooden floor from trunk section
3. ring 4m diameter
4. butterfly flower planting area
5. landscape area corten steel plate 12m diameter
6. secondary access with slate paving
7. seating area toward the ring with slate paving
8. gravel path
9. tree fixation of the ring
10. seating area toward the ring with slate paving and corten steel band

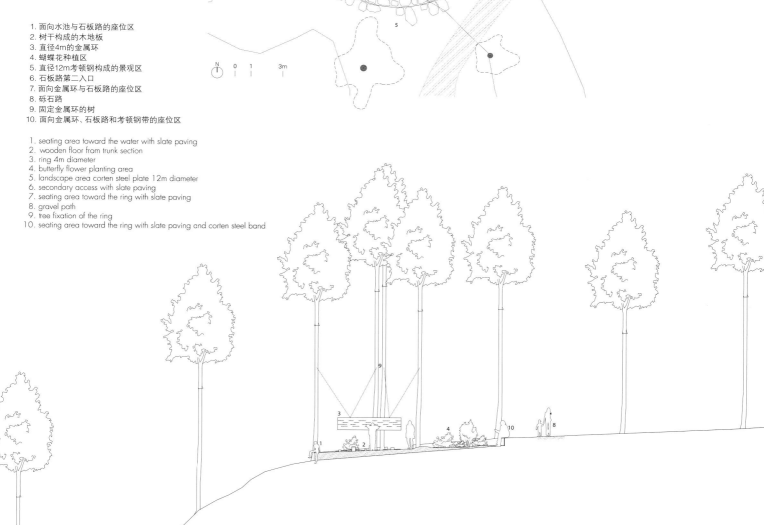

项目名称：Memorial at Utøya / 地点：Utøya island / 建筑师：3RW Arkitekter / 生态学家：Christian E. Mong / 竞赛图片制作：Fabian Schnuer Gohde / 施工场地：Volunteer parents of the victim and former members of the youth labour party (AUF - veterans) / 木地板工匠：Øystein Kjerpeseth, Nature AS / 结构工程师：NODE rådgivende ingeniører AS / 测量员：Fossum AS / 景观承包商：Anleggsgartner Finn Andersen / 钢制件（悬吊金属环）：Trosterud AS / 钢结构制作（景观环）：Vestre AS / 现场钢结构施工（景观环）：Skar Industriservice AS / 建筑现场管理：Telemark Vestfold Entreprenør AS / 项目管理：Jørgen Watne Frydnes, Utøya AS / 蝴蝶谷文：Dr. Joseph Chipperfield / 树木栽培家：Glen J. Read, Treesolutions AS / 类型：competition, 1st prize_commission / 状态：complete d on the 4th anniversary of the incident the 22nd July 2015 / 客户：AUF, Oslo / 功能：114m² of circular landscape intervention in a 12m diameter circle, including a suspended ring of 4m in diameter / 建筑面积：130m² / 设计时间：2014 / 竣工时间：2015 / 摄影师：©Martin Slottemo Lyngstad (courtesy of the architect)

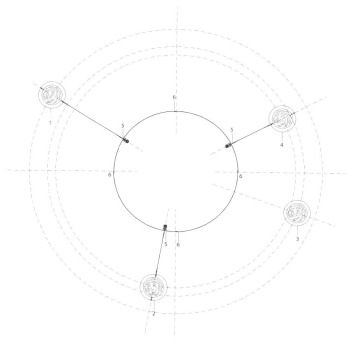

1. tree n°5
2. tree n°1
3. tree n°2 (optional)
4. tree n°3
5. fixation of two cables to the ring
6. welded fixation with bolts
7. grond level
8. 8mm steel plate
9. articulated fixation
10. welded plate
11. trunk perforation

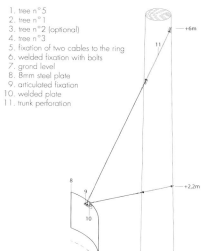

金属环与树木及连接缆线
ring with tree and cable connection

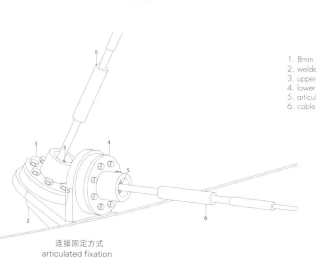

连接固定方式
articulated fixation

1. 8mm thick steel ring
2. welded plate
3. upper cable fixation
4. lower cable fixation
5. articulated rotule
6. cable connector

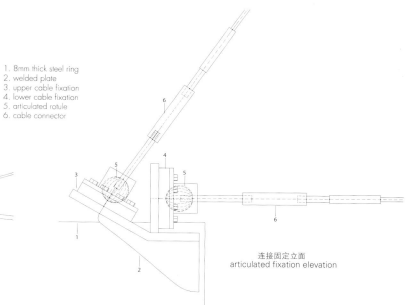

连接固定立面
articulated fixation elevation

>>154

Höweler + Yoon Architecture

Korean architect, designer and educator J. Meejin Yoon[right] studied at the Cornell University (B.Arch) and Harvard Graduate School of Design (M.Arch). Co-founded Höweler + Yoon Architecture LLP and MY Studio. Was awarded the Design Vanguard Award in 2007, Emerging Voices Award in 2007, and so on. Is currently Department Head of Massachusetts Institute of Technology. Eric Höweler[left] was born in Cali, Colombia and received B.Arch and M.Arch from the Cornell University. Is currently an Assistant Professor at the Harvard Graduate School of Design and principal of Höweler+Yoon Architecture LLP. Received Emerging Voices Award in 2007 and Annual Design Review from Architect Magazine in 2012.

>>12

Johan Celsing Arkitektkontor

Johan Celsing was born in Stockholm, Sweden in 1955. Studied architecture at The Royal Institute of Technology in Stockholm and painting at Academie de la Grande Chaumiere in Paris. Is professor of Advanced Design at the Royal Institute of Technology. Is a member of the Royal Swedish Academy of Fine Arts and Royal Swedish Academy of Sciences. Has lectured and acted critic widely at institutions such as: London Metropolitan University, Madrid ETSAM, Harvard GSD, Columbia University at New York, TU Münich, Royal Danish Academy of Fine Arts, Oslo School of Architecture. Major built work includes Museum Gustavianum at Uppsala, Millesgarden Art Gallery at Stockholm and Museum of Sketches at Lund. His work has had several nominations for the Mies van der Rohe Award. He has been awarded the Kasper Salin Prize two times.

>>34

Henning Larsen Architects

Is an international architecture company in Denmark, with strong Scandinavian roots. Founded by Henning Larsen in 1959, and is currently managed by CEO Mette Kynne Frandsen and Design Director Louis Becker. It has offices in Copenhagen, Oslo, Munich, Istanbul, Riyadh, the Faroe Islands and a newly established office in Hong Kong. Its goal is to create vibrant, sustainable buildings that reach beyond itself and become of durable value to the user and to the society and culture that they are built into. Its ideas are developed in close collaboration with the client, users and partners in order to achieve long-lasting buildings and reduced life-cycle costs. Won the prestigious European Union Prize for Contemporary Architecture-Mies van der Rohe Award 2013.

>>68
Ron Shenkin Studio
Tries to deliver a comprehensive design through the creative integration of innovation, creativity and originality. Its design solutions result from a collaborative process that refines the needs of customers and the project, together they create a program which serves as an anchor for the design. Ensures persist focus on details such as, accurate drawings and detailing using three-dimensional software. Integrates professional teams of contractors, professionals, and various suppliers to share their knowledge, together they deliver the highest standards.

>>58
DA Studios
Was founded in 2008 at the Hyderabad, India. Is owned and run by Dommu Krishna Chaitanya[left], Alluri Kasi Raju[right] and Srivalli Pradeepthi Ikkurthy[middle]. Krishna Chaitanya Dommu persued a graduate study of Technology in Architecture at the University of Nottingham, UK. Venkata Kasi Raju Alluri received his master in Advanced Architecture from the Institute for Advanced Architecture of Catalonia, Spain. Srivalli Pradeepthi Ikkurthy studied Histories and Theories of Architecture in her master course at the AA School of Architecture, London.

>>96
Passelac & Roques Architects
Romain Passelac is born in Carcassonne, Aude, France in 1978 and François Roques in Béziers, Hérault, France in 1977. They received Architecture Degree DPLG at Ecole Nationale Supérieure d'Architecture de Toulouse in 2002. After that, associated together their company in Narbonne, France in 2004.

>>138
SET Architects
Is based in Rome, Italy and led by 3 young architects Lorenzo Catena[left], Onorato di Manno[middle], and Andrea Tanci[right], specialized in designing at different scales, from interior to urban planning. The key in design procedure is in-depth theoretical research based on the investigation of the essence of architectural elements. The goal is 'space in direct dialogue with the people' through careful attention to context and local culture, experimental and coherent use of materials and through a critical approach to sustainability, and finally, by granting the architectural object as cultural factor. Was awarded the first prize with Bologna Shoah Memorial, Italy (2016).

>>162
Koishikawa Architects
Was founded at Tokyo, Japan by two Japanese architects, Hiroya Kobiki and Noritaka Ishikawa in 2011. Hiroya Kobiki was born in 1980, Kyoto and Noritaka Ishikawa was born in 1981, Tochigi. They graduated from the Kyoto Institute of Technology in 2006. Hiroya Kobiki has worked at Tomoko Taguchi Architect & Associates for 6 years. Noritaka Ishikawa has worked at Akira Kuryu Architects & Associates before founding Koishikawa Architects with Hiroya Kobiki.

>>96
Rudy Ricciotti
Was born in 1952 in Algeria and studied at Engineer school of Geneva and Architecture school of Marseille. Is one of the foremost representatives from the architects generation of combining creative prowess with a genuine constructive approach. Since he has been influenced by the Arte Povera and his buildings in the beginning of 1990s, has become more austere and functional, making use of minimalist and low-tech solutions. Pioneer and ambassador of concrete, Rudy Ricciotti sublimes innovative concretes in significant constructions such as the Museum of Civilisations of Europe and the Mediterranean in Marseille, the new wing for the Louvre in Paris to host its Islamic art collection, Jean Cocteau Museum in Menton.

Aldo Vanini
Practices in the fields of architecture and planning. Had many of his works published in various qualified international magazines. Is a member of regional and local government boards, involved in architectural and planning researches. One of his most important research interests is the conversion of abandoned mining sites in Sardini.

>>80
Hiroshi Nakamura & NAP
Was born in Tokyo, 1974. Has worked at the Kengo Kuma & Associates from 1997 to 2002. Established Hiroshi Nakamura & NAP Associates Inc. in 2002 Received numerous prizes at the JCD Design Award, Detail Prize, Good Design Award, JIA Award, AR+D Awards for Emerging Architecture, Green Good Design Award and LEAF Awards.

>>42
PLAN01
Is an organization of 10 partners from 5 associated agencies including Atelier du Pont, Ignacio Prego Architectures, Jean Bocabeille Architecte, Koz, and Philéas. Was founded in 2002 with a shared workplace in the 12th Arrondissment, Paris. They have amalgamated with the aim of creating a space where they can enjoy freedom, shielded from the particular constraints and demands that prevail in the architectural world.
At the same time, all the five firms work on their own individual projects. PLAN01 founded PLAN02 in 2008. It is an ecodesign office, works for the firms that make up PLAN.01.

>>126

Philippe Prost

Graduated from the Architecture school of Versailles with Architect DPLG in 1983. Received DESS (professional master) and DEA (research master) in Urban Planning at the University of Paris VIII. In 1989, got a DSA (specialized degree) from the Ecole de Chaillot, Paris.
Was nominated for the Equerre d'Argent by the Le Moniteur in 2009. Received Chevalier des arts et des lettres from French government in 2014. Is currently a member of Academie d'architecture and president of the National School of Architecture of Paris-Belleville.

©Giuseppe Gradella

>>172

3RW Arkitekter

Is located in Bergen, Norway and consists of 5 partners and 12 employed architects. Was established in the fall of 1999 as a young and independent office. Since then they have been working with a wide variety of projects and clients. At the moment they are engaged in projects expanding from singular housing in private context to various projects for local municipalities and companies designing public buildings, offices and urban housing densification. They focus on architecture as a tool for investigating social relationships within our society and consider themselves as a network-oriented company since it is their opinion that creating architecture, creating the framework of a dynamic society, is too big a task to be handled inside an office of architecture alone. Therefore they engage in many cooperative projects along with artists, scientists, engineers, anthropologists, ecologist, geographers and writers, in order to be a proper participant in the discussion of our surroundings. Several members of the office are also teaching at the Bergen School of Architecture (BAS).

>>116

Nizio Design International

Miroslaw Nizio studied at the Department of Interior Architecture and Sculpture, Academy of Fine Arts in Warsaw, and at the faculty of Interior Design, Fashion Institute of Technology in New York. He started his own design business in the 1990s in New York and opened Nizio Design International studio in Warsaw's Praga district in 2002. He is an architect who is mainly recognized for designs of public spaces including museums, historical exhibitions and commemoration monuments. The Company boasts long-standing experience of designing and completing architectural, revitalization, and exhibition projects. Nizio Design International has won the first award in the international competition for the design of the Wroclaw Contemporary Museum.

©2016大连理工大学出版社

图书在版编目(CIP)数据

建筑情感：从宗教到世俗：汉英对照 / 丹麦亨宁·拉森建筑事务所等编；杜丹等译. — 大连：大连理工大学出版社，2016.12
 ISBN 978-7-5685-0603-8

Ⅰ. ①建… Ⅱ. ①丹… ②杜… Ⅲ. ①丧葬建筑－建筑设计－研究－汉、英②纪念碑－建筑设计－研究－汉、英 Ⅳ. ①TU251

中国版本图书馆CIP数据核字(2016)第291298号

出版发行：大连理工大学出版社
　　　　　（地址：大连市软件园路80号　邮编：116023）
印　　　刷：上海锦良印刷厂
幅面尺寸：225mm×300mm
印　　　张：11.75
出版时间：2016年12月第1版
印刷时间：2016年12月第1次印刷
出 版 人：金英伟
统　　筹：房　磊
责任编辑：杨　丹
封面设计：王志峰
责任校对：周小红
书　　号：978-7-5685-0603-8
定　　价：228.00元

发　行：0411-84708842
传　真：0411-84701466
E-mail：12282980@qq.com
URL：http://www.dutp.cn

版权所有·侵权必究